THE COMMODIFICATION GAP

IJURR-SUSC Published Titles

THE COMMODIFICATION GAP

Gentrification and Public Policy in London, Berlin and St Petersburg

MATTHIAS BERNT

STUDIES IN URBAN
AND SOCIAL CHANGE
BOOK SERIES

Registered Offices
John Wiley & Sons, Inc., 111 River Street, Hoboken, NJ 07030, USA
John Wiley & Sons Ltd, The Atrium, Southern Gate, Chichester, West Sussex, PO19 8SQ, UK

Editorial Office
9600 Garsington Road, Oxford, OX4 2DQ, UK

For details of our global editorial offices, customer services, and more information about Wiley products visit us at www.wiley.com.

Wiley also publishes its books in a variety of electronic formats and by print-on-demand. Some content that appears in standard print versions of this book may not be available in other formats.

Library of Congress Cataloging-in-Publication Data

Name: Bernt, Matthias, author.
Title: The commodification gap : gentrification and public policy in
    London, Berlin and St. Petersburg / Matthias Bernt.
Description: Hoboken, NJ : Wiley, 2022. | Series: IJURR-SUSC | Includes
    bibliographical references and index.
Identifiers: LCCN 2021059922 (print) | LCCN 2021059923 (ebook) | ISBN
    9781119603047 (cloth) | ISBN 9781119603054 (paperback) | ISBN
    9781119603085 (adobe pdf) | ISBN 9781119603078 (epub)
Subjects: LCSH: Gentrification–England–London. |
    Gentrification–Germany–Berlin. | Gentrification–Russia
    (Federation–Saint Petersburg.
Classification: LCC HT170 .B47 2022 (print) | LCC HT170 (ebook) | DDC
    307.3/416–dc23/eng/20211217
LC record available at https://lccn.loc.gov/2021059922
LC ebook record available at https://lccn.loc.gov/2021059923

Cover Design: Wiley
Cover Images: © I Wei Huang/Shutterstock, © diesektion/Unsplash, © Pikoso.kz/Shutterstock

Set in 11/13pt Adobe Garamond by Straive, Pondicherry, India
Printed and bound by CPI Group (UK) Ltd, Croydon, CR0 4YY

C099050_130422

Contents

List of Figures

List of Tables

Series Editors' Preface

IJURR Studies in Urban and Social Change Book Series

The IJURR Studies in Urban and Social Change Book Series shares IJURR's commitments to critical, global and politically relevant analyses of our urban worlds. Books in this series bring forward innovative theoretical approaches and present rigorous empirical work, deepening understandings of urbanisation processes, but also advancing critical insights in support of political action and change. The Book Series Editors appreciate the theoretically eclectic nature of the field of urban studies. It is a strength that we embrace and encourage. The editors are particularly interested in the following issues:

- Comparative urbanism
- Diversity, difference and neighbourhood change
- Environmental sustainability
- Financialisation and gentrification
- Governance and politics
- International migration
- Inequalities
- Urban and environmental movements

The series is explicitly interdisciplinary; the editors judge books by their contribution to the field of critical urban studies rather than according to disciplinary origin. We are committed to publishing studies with themes and formats that reflect the many different voices and practices in the field of urban studies. Proposals may be submitted to editor in chief, Walter Nicholls (wnicholl@uci.edu), and further information about the series can be found at www.ijurr.org.

Walter Nicholls
Manuel Aalbers
Talja Blokland
Dorothee Brantz
Patrick Le Galès
Jenny Robinson

Preface

First and foremost, I wish to thank the five institutions and two particular individuals that were crucial for making this work possible. The Alexander von Humboldt foundation granted me a Feodor Lynen Stipend, which allowed me to dedicate my time to this project and to conduct empirical work abroad. This work would have been all but impossible without the patience and encouragement of my employer, the Leibniz Institute for Research on Society and Space (IRS), which granted me the time necessary to research and write this book and supported this detour from everyday business. Finally, and most importantly I wish to thank my two hosts in London and St Petersburg. Without the dedicated and continued support from Claire Colomb and the Bartlett School of Planning at the University College of London, the whole project would have never taken place. The same goes for Oleg Pachenkov and both the Centre for Independent Social Research (CISR) and the European University of St Petersburg. More than once, I was deeply impressed by the enthusiasm, reliability and imagination with which these magnificent colleagues supported my work. Both Claire and Oleg provided invaluable intellectual conversations and great company, which helped me along the cliffs of the project.

Numerous people assisted in the data collection and interviews or shaped the book by providing comments, ideas and background information in all three cities researched. I am very grateful to all of them. My debts here are too numerous to be listed in full. I extend thanks, in particular, to Michael Edwards, Michael Hebbert, Loretta Lees, Tim Butler, Chris Hamnett, Peter Williams, Alan Mace, Paul Watt, Duncan Bowie, Mike Raco, Jennifer Robinson, Hyun Bang Shin, Antoine Paccoud and all the other scholars who supported me in London.

My greatest thanks go also to Lilia Voronkova, Thomas Campbell, Dimitri Vorobyev, Irina Shirobokova, Katya Korableva and the other participants of the 'Research Laboratory', as well as to Anna Zhelnina and Konstantin Axënov, who all guided my way through St Petersburg. I am also grateful to Oleg Golubchikov for helpful comments on the matter of gentrification in Russia.

Berlin has been the place that has shaped my thinking about gentrification for a very long time. There have been so many people who have provided me with motivation, inspiration and information over the years that I don't know how to count them all. *Pars pro toto,* I would like to thank a few people specifically here. First and foremost is Andrej Holm, who has been a friend and a political and intellectual companion for decades. The same goes for Margit Mayer who has crucially influenced my thinking about cities since I was a student. Over the years, a couple of people have been especially important for my studies on Prenzlauer Berg, both by providing information and supporting my work, and also by disagreeing and discussing things with me. These include Carola Handwerg, Michail Nelken, Theo Winters, Ullrich Lautenschläger, Jochen Hucke, Ulf Heitmann, Hartmut Häußermann, Karin Baumert, Wolfgang Kil, and Wilhelm Fehse and Bernd Holtfreter (who both passed way too early).

Grateful appreciation is also due to Talja Blokland, Michael Gentile and Slavomira Ferenčuhová, who read earlier versions of individual chapters. Finally, Willem van der Zwaag, Mary Beth Wilson and Kerstin Wegel have provided superb technical support.

Above all, I thank Anja and Juri for their love, support and patience, which have allowed this book to be written.

Berlin, 23 October 2021

CHAPTER 1

Introduction

Gentrification Between Universality and Particularity

In the late 1990s, Prenzlauer Berg, a neighbourhood in East Berlin, experienced rapid changes. After decades of decay, more and more of its dilapidated residential buildings were bought up by investors, renovated and rented out with considerable price increases. Grey and weathered facades turned into colourful yellow, lilac and pale blue. The smell of coal produced by oven heating disappeared. The place became fashionable and more and more media reports came up with stories about Berlin's new 'in quarter'. Accompanying this, the composition of the population changed too, with newcomers tending to be younger, better educated and, as time went by, also richer than the established residents. Together with this new population came a wave of newly established bars, clubs, restaurants, boutiques, etc.

So far, the story hardly sounds spectacular – even readers who have never heard of Prenzlauer Berg will most likely be aware of similar changes in other neighbourhoods and cities. In fact, what happened in Prenzlauer Berg has been experienced in many places in the world, before and since. The term gentrification has now become the most common term used for this kind of urban transformation. The theme has become so omnipresent that hardly any international conference that is focused on the urban proceeds without presentations on gentrification. Stacks of books have been written on the subject and in many countries the term has entered everyday vocabulary. Gentrification, however, was a term invented by the British-German sociologist Ruth Glass, who described it as early as 1964, in the following words:

> One by one, many of the working class quarters of London have been invaded by the middle-classes – upper and lower. Shabby, modest mews

The Commodification Gap: Gentrification and Public Policy in London, Berlin and St Petersburg, First Edition. Matthias Bernt.
© 2022 John Wiley & Sons Ltd. Published 2022 by John Wiley & Sons Ltd.

and cottages – two rooms up, two down – have been taken over, when their leases have expired, and have become elegant, expensive residences. Larger Victorian houses, downgraded in an earlier or recent period – which were used as lodging houses or were otherwise in multiple occupation – have been upgraded once again. . . . Once this process of 'gentrification' starts in a district, it goes on rapidly until all or most of the original working class occupiers are displaced, and the whole social character of the district has changed.

(Glass 1964, pp. xviii–xix)

Changes from shabby and modest to elegant and expensive were, of course, exactly what activists in Prenzlauer Berg had in mind when they started using the term gentrification to depict the changes they saw happening in their neighbourhood in the late 1990s. They were met by sharp opposition. Public officials and urban planners, but also Berlin-based urban scholars, took the claim that the changes taking place in the quarter could be termed as gentrification as an insolent provocation (see Winters 1997; Häußermann and Kapphan 2002). Interestingly, the arguments brought forward at that time were neither confined to questions of data interpretation, nor did they move quickly to the possible implications for public policies and planning. More often than not, a rejection of the argument came together with de facto claims about the ontological status of the concept of gentrification. A larger group of critics posited that the concept of gentrification was developed in the contexts of the USA and the UK, with their untamed laissez-faire capitalism (and had limited value beyond these locations). European cities, in contrast – and Berlin, in particular – would be marked by stronger urban planning, more welfare state assistance and highly developed tenant rights that together would protect the city from the excessive development experienced elsewhere. Altogether, this would make the concept of gentrification inapplicable. A second group of critics addressed the situation from another direction. They argued that gentrification was a necessary and unavoidable companion of capitalist land and housing markets. As long as there was capitalism, there would be gentrification. Talking about rent regulations and planning strategies would, therefore, only turn attention away from problems that were systemic in nature.

Thus, whereas one line of critiques emphasised the particularities of Berlin and set them in contrast to a perceived Anglo-American 'normal', the other managed to do exactly the opposite and abandoned the specificities of Berlin to make a global critique of capitalism. Unwittingly, both perspectives made an age-old choice known from the field of comparative social research:

analysing the same situation and empirical data, the first argued in a clearly individualising way, whereas the latter rested on a universalising form of explanation. The outcome was diametrically opposed positions.

What both perspectives had in common, however, is that they effectively cushioned Berlin's planning and renewal policies against criticism. If planning, welfare protection and tenant rights were so strong that gentrification could not happen here, why change anything? If gentrification was so deeply embedded in the nature of global capitalism, why should one expect local policies to make a difference? In summary, while coming from opposite directions, both critiques effectively shielded the policies of 'Careful Urban Renewal' exercised in Berlin at those times (see Chapter 5 for details) against criticism and helped in defending the status quo.

With this study, I want to suggest a different perspective on gentrification. I will show that gentrification is, indeed, a universal phenomenon that reflects general conditions set by capitalist land and housing markets, yet at the same time, it is only made possible through specific institutional constellations. Gentrification, I argue, is at the same time economically and politically determined. It rests on historically specific entanglements of markets and states, expressed in multiple combinations of commodification and decommodification. Analysing the historically specific nexus between commodification and decommodification in driving gentrification is, therefore, central to this book.

What is meant by decommodification and commodification? Under capitalism, most housing is produced for the purpose of being sold as a commodity in the market. At the same time, housing is an essential human need. Most societies have, therefore, found ways in which the production and/or consumption of housing is completely or partly sheltered against the markets, so that its character as a commodity is limited and/or restricted. Thereby, commodification and decommodification stand in a dialectical relationship. Commodification happens when the social use of housing is subordinated to its economic value. When housing is commodified, it can be treated as an investment and can be purchased, sold, mortgaged, securitised and traded in markets. Decommodification occurs when exactly the opposite is taking place (see Esping-Andersen 1990, p. 22). When the provision of housing is rendered as a right and/or when a person can maintain accommodation without reliance on the market, or when the conditions in the markets make it impossible to trade housing or invest in it, the commodity status is loosened and housing becomes decommodified.

With this study, I argue that it is only when decommodification is limited to a degree that allows for satisfactory rates of return on investment in housing that upgrading becomes lucrative for investors and gentrification is achieved. This is what I call the 'commodification gap'. I argue that it lies at the heart of what makes a difference when comparing gentrification across varied contexts. The major point here is that the general dynamics of commodification are universal in capitalist societies, whereas the ways in which markets are embedded into societies and the variations in which social rights are perceived, negotiated and legislated are not.

The treatment of gentrification as either a general feature of capitalism or a local experience – as argued in the debates in Prenzlauer Berg that I have described – thus, rests on a false dichotomy. However, this treatment is not unique to Berlin. Quite to the contrary, the difficulty of balancing historical and geographic particularities with the status of gentrification as a general theory has been a problem that has occupied urban scholars for a long time. Gentrification studies have been characterised by a progressing seesaw motion between universalising and individualising approaches for decades. When the term gentrification was invented by Ruth Glass, it primarily reflected on a very British (and to some degree even London-based) experience. Picking up on historical English class structures, the term gentrification had both a descriptive and an analytical edge, but always with a close connection to London. This is also reflected in the title of the book: *London: Aspects of Change*. Only a few years after the development of the term, it travelled to the other side of the Atlantic and stimulated a first wave of studies in North America. Here, the background was a novel and counter-intuitive 'back to the city' movement of middle-class households experienced after decades of 'urban crisis' and 'white flight', for which new explanations were needed. In this context, gentrification appeared as the term du jour to describe a new phenomenon. By and large, two major forces were seen as driving it – and both seemed to be of universal value (at least in Canada and the USA): (i) the sociocultural transformations accompanying the dawn of a post-industrial society, resulting in the rise of a new middle-class and (ii) the discovery of inner cities as a renewed terrain for investment strategies. Both transformations made up the core of explanations for gentrification back then and for more than two decades 'gentrification debates battled back and forth over the "post-industrial, new middle-class thesis" and the "rent gap exploitation thesis" over what had caused the rise of gentrification' (Shin and López-Morales 2018, p. 15). While gentrification expanded fast as a research field and became a major battleground of scholarly debates, differences between the place of its origin (London) and cities like Philadelphia, New York, Vancouver and others were hardly raised as a matter of concern.

In a second wave, the gentrification concept travelled back and forth across the English Channel to Western Europe, i.e. to a context that had also experienced deindustrialisation, suburbanisation and the growth of a service economy, but with some delay and a different history of urbanisation and housing. Here, too, empirical studies (by and large) stayed with the concepts imported from the USA/UK. At the same time, the particularities of West European cities, as opposed to those in the UK and the USA, were raised as issues of concern and attributed to 'contextual factors' or 'modifications'. If the first phase of gentrification studies had been uninterested in the issue of comparability, the contributions in this second phase were marked by attempts to contextualise what was still seen as a global phenomenon explained by universally valid theories.

From the late 1990s, this mood changed considerably. In the USA and the UK, gentrification had become such a cottage industry of academic career building that not only were more and more studies produced, but also more and more phenomena that were not included in the classical gentrification canon were now characterised as gentrification. Research emerged about commercial gentrification (Bridge and Dowling 2001), new-build gentrification (Davidson and Lees 2005, 2010), tourist gentrification (Gotham 2005), rural gentrification (Phillips 1993, 2004), studentification (Smith and Holt 2007), super-gentrification (Lees 2003; Butler and Lees 2006) and other forms of upgrading that did not follow the traditional demographic, cultural, and spatial patterns known from earlier gentrification studies. This expansion of gentrification research led to a growing sense of frustration among urban scholars, and more and more academics came to see the concept as overstretched[1] (see Bondi 1999; Lambert 2002; Hamnett 2003; Butler 2007; Maloutas 2007, 2012, 2018).

Where the concept of gentrification was applied outside its places of origin, critiques were even stronger. Thomas Maloutas argued (in the context of Athens) that gentrification was 'a concept highly dependent on contextual causality and its generalised use will not remove its contextual attachment to the Anglo-American metropolis' (Maloutas 2012, p. 33). The concept, he claimed, would be 'detrimental to analysis, especially when applied to contexts different from those it was coined in/for' (Maloutas 2012, p. 44). As a consequence, Maloutas demanded that the concept of gentrification not be used outside the Anglo-Saxon world and that more localised concepts and descriptions be found and used.

With the advancement of postcolonial approaches, these concerns have gained growing acceptance in the subsequent decade and are now widespread in the field of urban studies.[2] Nowadays, more and more scholars tend to see gentrification as an urban phenomenon rooted in very specific experiences realised in a handful of Western metropolises in the second half of the twentieth century (see Bernt 2016a). To an increasing extent, the concept is portrayed as overstretched (Schmid et al. 2018) and blamed for oversimplifying essentially variegated urban experiences by 'blindly apply(ing) theories from the West' (Tang 2017, p. 497). Some authors have even gone so far as to claim that the concept of gentrification has 'displaced and erased alternative idioms and concepts that may be more useful for describing and analysing local processes of urban change' (Smart and Smart 2017, p. 519) and have suggested that the concept of gentrification 'should be laid to bed . . . among those 20th century concepts we once used' (Ghertner 2015, p. 552). In this view, gentrification is successively seen as a thin theory, considerably over-stretched and not capable of integrating new developments.

While doubts and reservations about the concept of gentrification have become stronger than ever before, the last few years have also seen a remark-able expansion of empirical research on the subject carried out on an increas-ingly global scale. These days, scientific papers on gentrification come from all continents, including places as different as Yerevan (Gentile et al. 2015), Mexico (Delgadillo 2015), Copenhagen (Gutzon Larsen and Lund Han-sen 2008), Manila (Choi 2016) and many more. A research group around Loretta Lees, Hyun Bang Shin and Ernesto López Morales has alone pro-duced two volumes and two special issues that have applied the concept of gentrification to empirically studying many places around the globe (Lees et al. 2015, 2016; López-Morales et al. 2016; Shin et al. 2016). Building upon this work, Lees et al. argue that gentrification has indeed become a 'planetary process', and defend the application of the gentrification concept as follows:

> We have considered whether the concept of gentrification has global application, and whether there really is such a thing as gentrification generalized. After much research and international discussion . . . we have concluded YES – as long as we keep gentrification general enough to facilitate universality while providing the flexibility to accommodate changing conditions and local circumstances.
>
> (Lees et al. 2016, p. 203)

In summary, it can be said that today the urban studies world in gen-eral is split by a schism between the custodians of general (usually Marxist) theories and those criticising their Eurocentric implications. Yet, there are also voices that break out of this either-or way of arguing. Applying the con-cept of gentrification to cities in the Global South, researchers have found new ways of productively working with differences (for an overview see Lees et al. 2015; Shin and López-Morales 2018; Valle 2021) in the subsequent years, which have opened up new perspectives on the debate. Two issues stand out here:

First, manifold contributions have highlighted that capitalist markets in land and housing are limited in many cities in the Global South. Instead, hybrid arrangements of property rights are common (Lemanski 2014) and large parts of the housing stock are managed as informal housing (Ghertner 2014; Doshi 2013, 2015; Cummings 2016; Gillespie 2020). In effect, customary and state land tenures instead of private property are the rule, rather than an exception, in many countries. As a consequence, the movement of capital is limited in much of the Global South and gentrification is only possible when these limits are removed. The simultaneous existence of commodified and non-commodified land has given rise to a 'real estate frontier', i.e. an 'interface between real estate capital and non-commodified land' (Gillespie 2020, p. 612), which advances through the enclosure of uncommodified land. Other than in the Global North, market-based forms of land allocation cannot be taken for granted here, but need to be actively produced. Non-commodification, or de-commodification, is central here, and gentrification is only brought into being through a process of 'primitive accumulation' in which non-capitalist forms of tenure are actively attacked and abolished.

Closely related, research on gentrification in the Global South has repeatedly emphasised the centrality of state agency and extra-economic violence as a driver for gentrification (see Islam and Sakızlıoğlu 2015; Shin and Kim 2016; Shin and López-Morales 2018; Valle 2021. This argument has most thoroughly been developed by the geographer Gavin Shatkin (2017), who depicts how state interests are at the core of land monetisation in Asia. While Shatkin finds the specifics of gentrification theory (as formulated by Smith 1979 and others) of limited relevance for his cases, he still insists that it can provide useful insights. For the case of Southeast Asia, Shatkin argues that state actors are the main players in the real estate market. They use land rent capture for their own empowerment, as a source of revenue for the state, or redistribute the profits to key allies of the ruling elites. In turn, state actors become interested in maximising their control over land markets and exploiting the rent gap to its maximum. What makes this conceptualisation interesting is that it dissociates gentrification from its attachment to a specific geographic and socioeconomic context and uses the rent gap in a purely analytical form. For Shatkin, gentrification is not of interest per se, but rather one of a number of concepts combined to explain the complex development of state–business relations in real estate development in Asia.

This treatment resonates well with the discussion of a 'triangular entanglement of state, market, and society' described as the driver of 'three waves of state-led gentrification' investigated in China (He 2019). In her work, He describes how 'gentrification has become an integral part of the making of the modern competitive state in China' and analyses how 'the state's endeavour in extracting values from land/housing redevelopment [operate] through market operation' (He 2019, p. 33). Contrary to Western ideas about the predominance of markets in allocating land, He insists on including the developmentalist state in the analysis in China and emphasises that markets can only be understood as working in tandem with the state and societal forces. This, in turn, makes visible the limits of a restrictively economic explanation of gentrification that focuses mainly on the effect that markets play.

Summing up, the perspectives on the usefulness of the term gentrification in the urban studies community are more controversial than ever before. After half a century of research, gentrification is intensively brought into question as a concept today and the scientific community finds itself split into two camps. On the one hand, many scholars attack the 'diffusionist' practice of exporting the Western concept of gentrification to contexts where it is not seen as applicable. They claim that gentrification has been overstretched as a theory and has become a Procrustean bed for the analysis of essentially different urban experiences. On the other hand, we find academics ferociously defending the usefulness of the concept, calling for more flexibility and attacking what they see as 'fossilization, rather than contextualization' (Lees et al. 2016, p. 7). Third, we find contributions (positioned at the margins of this debate) that offer a third way to solve the *problematique* by decontextualising gentrification theory (see also Krijnen 2018) and using it as a conceptual device to be combined with other instruments.

This book aims to advance this debate beyond the dichotomist treatment of the 'universality vs. particularity' binary. The major theoretical proposition is that land rent capture and capital accumulation and, thus, gentrification can *always* and *everywhere* only be understood as embedded in specific institutional contexts. Institutional contexts and economic dynamics can, therefore, not be separated but, instead, need to be integrated into the analysis.

Against this background, the book at hand provides an attempt to put the 'state question' at the centre of the explanation and rethink the relationship between markets and states in the field of gentrification. It does so through three empirical case studies that lay out how this relationship has developed in three neighbourhoods located in different countries, and uses this material to produce a novel concept.

How to Compare? Why Compare?

The ambition of this book is thus directed towards theory building, which will be achieved through a comparison of empirical materials. But how can gentrifications happening in different cities and at different times be meaningfully compared? Answering this question demands some reflection on the ways in which comparisons are composed, why they are done and how the similarities and differences observed can be brought together. Against this background, this chapter now lays out the basic etymological and ontological orientations of this study and presents its methodological design.

Separating the shared essence of a phenomenon from its various expressions is an epistemological problem that puzzled even ancient philosophers. It has proven to be so fundamental to our understanding of the world that wrestling with it has resulted in fundamentally different axiomatic positions and practices in scientific research. In this sense, the difficulty of coming to grips with the applicability of the gentrification concept beyond the classical cases of the UK and USA described above reflects a deep-seated problem faced by social science in general.

Crucially, in conceptualising gentrification as either a universal phenomenon or a specific local experience only relevant to a handful of cities, the proponents of the described debate have entered the rocky waters of comparative methodologies. While doing so, they have necessarily taken on board a series of long-established methodological problems that come along with the comparison of complex social phenomena.

The first is the use of comparison. Why are comparisons done? What is the point of comparing urban change in New York's Lower East Side in the 1980s with something that is going on in, say, Bangalore today? In fact, there are many reasons for carrying out a comparative study. The aim could be to explore whether a theory developed about the causes of gentrification (e.g. middle-class invasion, the rent gap, or cultural upgrading) holds true when some of the variables it is built upon vary. Alternatively, one could examine a small number of cases holistically to see whether similarities or differences observed between the cases can be related to causal conditions. These two ways of proceeding can be termed as variable-oriented vs. case-oriented strategies of comparison (Ragin 1987, pp. 54–55). Some authors have also argued for comparison as a mode of thought, enabling 'defamiliarisation' and assisting in uncovering the hidden assumptions a theory is built upon (Robinson 2006; McFarlane 2010). In this view, studying a process in a non-familiar environment (e.g. gentrification in a shrinking city) could function as an eye-opener and allow factors and connections that are hidden elsewhere to be revealed.

No matter the particular motivation, the essential goal of all comparison is not to describe something, but to understand and explain it. The point here is to ask the reasons why a particular comparison is made. If apples and bananas are compared with donkeys, we need to have an idea about what constitutes fruits in contrast to animals. Yet, what exactly is the link between a theory, say about fruits, and the need to compare? While most comparative researchers agree that comparisons refer to theory, there is no privileged stage of theory building in which it is most appropriate. Comparing can, thus, contribute to uncovering the limits of a theory, it can help to draw a contrast between cases and it can be used to suggest testable hypotheses. These hypotheses, in turn, can create demand for a new theory that can then be tested again by including new cases (Skopcol and Somers 1980). The relation between theory building and empirical comparative work is, thus, cyclical and there is no hierarchical (but rather a mutually reinforcing) relationship between theory and comparative empirical observation.

The current treatment of gentrification in urban studies as either a particular local experience, or a planetary urban phenomenon, occupies an uneasy position when examined against these potentials (see also Bernt 2016a). While individualising accounts have proven to work well for attacking the usefulness of gentrification theory for a specific case, they more often than not leave one wondering whether the problem is just a misclassification of an individual case (e.g. when the gentrification concept is applied to cases where gentrification doesn't happen), a linguistic issue or something that is more deep-seated and inherent in the concept of gentrification. Universalising accounts, on the contrary, often relegate differences to the role of contextual factors, local specificities or contingencies, seen as negligible for theory building. Here, the consequence is an immunisation of established theories, but hardly their advancement. In sum, both ways of comparison fall short. This results in a stalemate in which both sides employ more or less convincing evidence to support their claims, but leave existing theories untouched and fail to frame a way forward.

One reason for this lies in the particular limits that are specific to either of these forms of comparison. There have been elaborate discussions about the difficulties of comparing complex social phenomena (like revolutions or cities) that are characterised by 'small n, many variables' (Lijpart 1971) in the past, especially within the field of comparative history. As a consequence, manifold useful strategies for conducting comparisons have been proposed (see for example Przeworski and Teune 1970; Lijphart 1971; Sartori 1970, 1991; Smelser 1976; Tilly 1984; Skocpol 1979; Skocpol and Somers 1980; Ragin 1987). In this context, Charles Tilly (1984) has developed a widely cited typology that distinguishes four types of comparison along the two dimensions of scope and number (see Figure 1.1). Thereby, in scope, comparisons can range from quite particular (getting the case right) to quite general (getting the characteristics of all cases right) and in number, comparisons can range from single to multiple.

On this basis, Tilly (1984, pp. 82f.) distinguished between individualising, universalising, generalising (or variation finding) and encompassing comparison. In this scheme, individualising and universalising comparisons

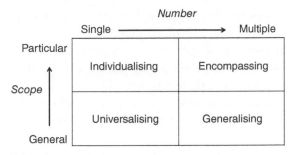

FIGURE 1.1 Types of comparison (Tilly 1984).

are antidotes: individualising comparisons contrast specific instances of a phenomenon as a means of grasping the particularities of each case, whereas universalising comparisons aim to establish that each instance of a phenomenon follows the same rule (ibid.). The point is that both types of comparison can be made, but each has a different function for different purposes. Individualising comparisons work well to illustrate a given theory. They can also be very effectively used to challenge the validity of a theory. The problem, however, is that they leave much to be desired when it comes to developing a new theory. Universalising comparisons, in contrast, are an effective means to go beyond the surface of similarities and dissimilarities and produce models and generalisations that allow for theorising. More often than not, the price for this is a neglect of differences.

Expressed differently, both types of comparison have their particular strengths and weaknesses – but only in relation to a specific strategy in making a theoretical argument. What counts is not the comparison as such, but the theoretical argument. It is in this field that both individualising and universalising approaches have their Achilles' heel.

Concepts and Causation

Necessarily, this leads us to matters of causation and explanation. How can empirical facts be connected to theory? How can we determine that something we can see and document can meaningfully be called 'gentrification'? While the popular image of gentrification, reflected in a now globally popular blend of artists and cappuccino places, renovation activities and rising housing costs, seems to suggest an easily recognisable picture on the surface; however, the underlying substance of gentrification is more difficult to track. The reason for this is the fundamental philosophical difference between the essence of a phenomenon, its manifestation and its description. For this reason, this section will discuss ontological and epistemological problems of causation and conceptualisation, thereby sharpening our understanding of the relationships between concrete phenomena, abstraction and causation, as well as laying out some basic ideas about the value and the pitfalls of connecting empirical phenomena and theory. In what follows, I base my arguments on what has become known as 'critical realism' in the philosophy of social sciences (Bhaskar 1975; Harré 1986; Stones 1996; Yeung 1997; Sayer 1992, 2000).

One of the most common fallacies to be found in gentrification research is the treatment of gentrification as a 'real' phenomenon, instead of as a concept. More often than not, one encounters phrases like 'gentrification has become a global phenomenon', 'gentrification has expanded beyond its places of origin', 'gentrification has reached into new neighbourhoods', etc. What these phrases have in common is that they reify gentrification. They treat gentrification as something that has an equivalent in reality, i.e. as an objective fact that can be measured, described and assessed visually. If one examines the definition(s) of gentrification more closely, however, it soon becomes clear that what is usually referred to as gentrification is a bundle of empirically observable phenomena, rather than a singular object. Most importantly, these are:

1. an immigration of middle-class households into areas where they were not prevalent before;
2. investments to upgrade houses and infrastructure (e.g. shops, or restaurants);
3. purchases or rentals of homes by more affluent buyers/renters;
4. rising house prices or rents;
5. a decline in the number of working-class and other low-income groups living in the area; and
6. a change of the social character of a neighbourhood.

In terms of observation, these are distinct empirical objects. What gentrification theories do is bundle together these objects under the term gentrification. The problem is that all these properties are also linked to social relations outside of gentrification. The immigration of middle-class households has, thus, been linked to broader social, cultural and demographic shifts in Western societies (Ley 1996); the investment of money into housing has been connected to the growing financialisation (Aalbers 2012, 2016; Haila 2016) of the real estate sector; and the decline of working class housing occupants has been explained as an outcome of the professionalisation of occupational positions in 'knowledge economies' (Hamnett 2003), etc. Thus, what is usually coined gentrification is not only a wide umbrella, but the phenomena covered by it are co-determined by factors that stand outside neighbourhood change. It is, thus, easily understandable why gentrification has been termed a 'chaotic concept' (Beauregard 1986) that lumps together unrelated phenomena and diverse explanations.

But what then is the ontological status of gentrification? What is the relationship of empirically observable phenomena to the concept? Different ontological positions allow different answers. Thus, from a positivist point of view, a bundle of empirical phenomena could be labelled gentrification as long as the term accurately describes patterns of similarity and regularity with which phenomenon A (say the renovation of dilapidated houses) is followed by phenomenon B (say a decline in the number of poor households). From a hermeneutic standpoint, the term gentrification would make sense only in so far as it was able to reveal the meanings that are attributed to the actual actions underpinning the process. Calling something gentrification would, thus, be adequate when the term could be found in actual discourses and used by actual actors, e.g. to demarcate social demands discursively. A realist understanding of the question would look rather different and, under this view, putting together phenomenon A (a renovation of dilapidated houses) and B (displacement of poor households) under a concept X (like gentrification) would only be justified if one was able to establish a causal relation between the two. The major consequence of a realist approach to gentrification is that conceptualisation, not research methods, is put at the centre.

Yet, how then can causality be built and examined? At the most fundamental level, each concept or theory is an abstraction. It must abstract from the particular conditions of a specific case and select those constituents of the case that have a significant effect that might also be found in other cases. In other words, the multiple elements of a concrete object must be isolated from each other, examined one by one and then recombined to form concepts that reflect the concreteness of their objects. The difficult point about abstractions is that they require the separating out of the kind of relationship that a particular element has to the concrete totality of which it is a part. In this respect, it is crucial to distinguish between external or contingent relations and internal or necessary relations (Sayer 1992, pp. 89–90). External or contingent relations can be found when two objects can exist without each other. Sayer (ibid.) uses the example of a person and a piece of the earth for this; though their interference can have significant effects, each of the two can exist without the other. By contrast, the relationship between a master and a slave or a tenant and a landlord is internal: both entities form a social relation that ceases to exist without both elements being part of it. Third, it is possible to identify asymmetrical internal relations. These appear when 'one object in a relation can exist without the other, but not vice versa' (Sayer 1992, p. 90). Sayer provides the example of the state and council housing; states can exist without council housing, but no council housing can exist without a state.

Clarifying the type of relationships that link the different elements of an object to each other is the first step towards conceptualising, as it forces the researcher to explain what it is about an element that makes it a necessary part of a relationship. With regard to gentrification, the questions would thus be: What is it about phenomenon A (renovations of dilapidated houses) that enables it to cause phenomenon B (displacement of poor households)? Why and how are renovations done and how does this relate to the likelihood of poor households being displaced? Is there a phenomenon C (say an immigration of middle-class households) that necessarily goes together with phenomenon A to achieve phenomenon B?

In this sense, causation is the identification of the actual mechanisms producing the event (as different from identifying patterns of regularity). This, however, leads to new problems as the 'open systems' character of the social world leads to situations in which the same causal power can produce different outcomes in different contexts, or different causal mechanisms can work at the same time. As multiple causation is fundamental to social phenomena, it follows that a particular event (e.g. the displacement of low-income households) can involve several causal mechanisms that may or may not internally be related to each other. Depending on the conditions, the same mechanism can produce different results, or different mechanisms can lead to the same outcome. If one was to relate this approach to the example of gentrification, the research design would become even more complex: not only would a causal relation between A (renovations of dilapidated houses) and B (displacement of poor households) need to be established and the impact of the intervening variable C (immigration of middle-class households) be explained, but research would need to expand to other contexts in which the impact of C (immigration of middle-class households) would possibly be neutralised by variable D (say strong welfare support) or intensified by E (say ethnic discrimination). From this, theorising would move backwards and re-examine the causal relationship between renovation activities and displacement established at an earlier stage of research and reformulate the theory on this basis.

The realist approach towards theory building is, thus, built upon empirical observation and abstraction. It starts with a theory, but moves on to empirical data capable of inducing an alternative conceptualisation, and then moves back to revising the original theory, thus producing a need for new empirical observation and so on and so forth. The actual processing of research is, therefore, highly iterative. It implies a back and forth between inductive and deductive cycles of data collection and explanation in which deduction helps to identify the object of interest and suggests explanations about the mechanisms at play, whereas induction provides data that is to be explained and tested. Both the use of data as an illustration of theoretical arguments and the use of empirical data to falsify a theory without giving any specificities that would allow for alternative conceptualisations are seen as preliminary stages in the development of new knowledge that need to be overcome by the development of new theories.

Design of this Study

How can these ideas inform a comparative study on gentrification? Without further ado, I draw four basic orientations from the preceding discussion.

First, I conceptualise gentrification not as an object, but as an abstraction. Advancing its study is hardly possible on the basis of a holistic approach that tries to integrate a broad variety of empirically observable phenomena all at once and indiscriminately apply the whole range of available, and to some part contradictory, explanations for gentrification. Learning from realist approaches, it is necessary to disaggregate the various explanations and start with one established theory whose validity can be examined in different contexts. Only when this is done, does it become possible to carefully examine the limits of this theory of gentrification and to study how other factors impact on the causal relationship between different phenomena. Meaningfully studying gentrification, thus, demands a clear point of reference towards an existing theory, as well as a focus on the phenomena to be observed.

Second, there is a close relationship between studying how an established concept works in different contexts and theoretical advancement. The task of an empirical study is, thereby, not to verify or falsify an established theory, but to work out alternative causations. In this context, it becomes crucial to empirically distinguish between necessary and contingent relations. Do different structures (say welfare states) influence a mechanism in a way that changes the causal relationship between two phenomena (say renovations and displacement)? Do contextual factors lead to differences in kind or in degree? Can the same outcomes (e.g. displacement) be explained by different mechanisms in different environments?

Third, neither universalising nor individualising modes of comparisons suffice to promote this process of iterative theory development. The problem, as I have pointed out with reference to Tilly, is not that particularising or universalising comparisons were misleading, as such, but that they are good in doing different things. While universalising comparisons work well to support a theory, individualising comparisons do exactly the opposite, demonstrating where a theory has reached its limits. The benefits and drawbacks of the two types of comparison are laterally inverted. What is needed is not a decision for the Scylla of universally valid categorisations or the Charybdis of thick descriptions of singular cases, but a way between the two that engages established theories with new empirical material that aims to build better explanations. In this sense, comparisons can be generative for a new round of theorising – but only when they are built upon a reference to existing theories and a commitment towards abstraction. Working with concepts is of key importance here.

Fourth, the relationship of theorising and empirical research in the course of investigation can be described as 'ascending from the abstract to the concrete', as Karl Marx famously put it in *Grundrisse* (1971 [1894). The abstract defines the object of empirical research that then leads to understanding the points where the theory stops explaining its particular objects and new theories must be integrated or found. This leads to an iterative back and forth movement between the empirical observation, abstraction and conceptualisation that is illustrated in Figure 1.2.

The study starts with a simple, abstract concept that seems to apply to all sorts of gentrifications at all times; moves on to the examination of actually existing forms of this abstraction; examines the interrelation of this abstraction with different concrete sociospatial formations; and then moves back to the abstract and examines the dialectic developed at a systemic level so far.

How can these recommendations be taken on board? In essence, restriction is demanded both with regard to the issues studied and the theories examined. In this respect, the following points form critical junctures for this book.

1. *Definition of gentrification:* I follow Clark's (2005) definition of gentrification and count all processes as gentrification that 'involv[es] a change in the population of land users such that the new users are of higher socio-economic status than the previous users, together with an associated

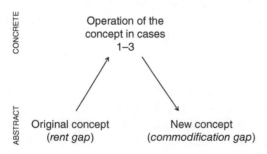

FIGURE 1.2 The relationship of abstraction and empirical observation in this study.

change in the built environment through a reinvestment of fixed capital'
(Clark 2005, p. 258). With this, I position myself within close range to
what has become known as 'supply-side theories' in the world of gentrification researchers.[3]

Two empirically observable issues are the focus of my interest.
These are: (i) the investment of capital into the built environment (see
Smith 1979, 1996) and (ii) the displacement of low-income residents by
more affluent users (see also Marcuse 1986). In my view, these are the basic
conditions that make gentrification different from other forms of neighbourhood change, as well as the two prerequisites without which the concept of gentrification would cease to make sense. Achieving higher quality
and returns on existing pieces of land and housing is (in the long run) not
possible without putting work and capital into it. Capital, in turn, is only
invested when a satisfactory rate of return is achieved that is higher than
what is already gained from the piece of land in question. Gentrification
and investment are, thus, causally connected. In a similar vein, gentrification is all but impossible without the exclusion of groups of people. If
everybody was able to use land, housing and infrastructure in the same
way in a particular area, displacement could not occur and would either
become pure coincidence or be caused by something other than gentrification.[4] For all these reasons, reinvestment and displacement form the two
key variables that are examined in all the cases studied in this research.

2. *Structuralism as a starting point:* The rent gap theory provided by Neil
   Smith (1979) builds the theoretical point of reference for this study. The
   main reason for this is that the theoretical ambitions of supply-side theories have always been more universal than those of their demand-side
   counterparts. Arguably this makes them a more attractive starting point
   for a comparative study. The goal is, thereby, not to verify or falsify
   supply-side theories, but to examine how theories work in different contexts, identify their limits and suggest new theories.

3. *Methodological   substantivism,   institutionalism   and   historicism:*
   Gentrification is not analysed in a formalised way as an expression of
   abstract political economies, but it is situated in space and time. Thereby,
   institutions are given a primary role in the explanation of gentrification.
   These are understood as historically and geographically specific sets of
   formal rules, compliance procedures and standard operating practices
   that structure the relationships and the interactions between individuals
   (see Hall 1986, p.19). As such, they have more formal status than cultural norms, but they are not necessarily legally codified. With this definition, there is a wide variety of institutions, legal codes, administrative
   organisations, customary habits and traditions. This study, however,

focuses on formal institutions – like forms of tenure, property regulations and rent laws. This choice is motivated by the cases analysed, in which formal institutions play a central role. It should not be seen as conceptually exclusive; depending on the context, other institutions (e.g. caste, family, ethnicity) might be more important.

4. *Comparitivism:* The project applies a comparative research strategy and examines heterodox cases of gentrification. This strategy is, however, distinguished from formal and static ways of comparing distinct cases, as it seeks to engage observable differences between the cases for the purpose of broader theorising. In order to stretch the boundaries of imagination and to 'provincialize' (Chakrabarty 2000) ill-informed universalisation, it actively searches for dissimilar cases and seeks to interrogate different modes of interaction between the market and state in terms of gentrification. Heterogeneity is, thus, sought as a means to identify diverse combinations of factors.

The operational model of this research is the following: reinvestment is thought to result in displacement (which is the core of gentrification), while at the same time different institutions can impact on this interrelation in a way that can modify, alter and even prevent both reinvestment and displacement from happening. As a consequence, different forms of gentrification can emerge. The interrelation of these elements is expressed in Figure 1.3.

The empirical backbone of my argument stems from three empirical studies in London, Berlin and St Petersburg. As the universal conditions enabling gentrification, 'namely, rent production, reproduction, and capture; production of gentrifiers and gentrifiable areas; and class displacement/ replacement' (Betancur 2014) are present in all three cities, one can assume it will be possible to find gentrification in each of them as well. However, as I will

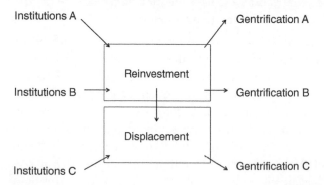

FIGURE 1.3 The relation of institutions, reinvestment, displacement and gentrification.

show, although instances of gentrification can be observed in all three cases, their dynamics, spatial patterns and causalities are highly diverse. In other words, I study a sample of dissimilar gentrifications, and while the cases studied support the idea that gentrification is a somewhat global phenomenon, they also demonstrate how this global phenomenon operates in fairly varied ways and is based upon different sets of local and national relations.

Having manifold representations of gentrification in mind, I focused on inner city residential neighbourhood change. I thus went to the 'usual suspects' of gentrification. The reason for this is that I aimed to include urban environments that would be connected to gentrification by most observers. Moreover, two of the cases selected have a longer history, allowing us to study the impact of changing institutions and regulatory environments over time. Typically, such cases are to be found in inner city environments. Based on these considerations, I selected Barnsbury (London) and Prenzlauer Berg (Berlin) as case studies. Both these neighbourhoods are seen as classical cases in their contexts and have been gentrified over decades. Case selection was more difficult in Russia, as gentrification is a new phenomenon there and often seen as a matter of newly built gated communities on city fringes or in former industrial wastelands, rather than as an inner city phenomenon. Nevertheless, I studied gentrification in the 'classical' environment provided by St Petersburg's inner city (Tsentralnyi Rayon). While this might be seen as being prone to the fallacy of morphological determinism, I argue that it is exactly the difficulties that gentrification faces in these environments that provide lessons to be learned. In some respects, the case of St Petersburg can be regarded as a control case, allowing insights that cannot be gained in Germany or the UK.

Why did I study gentrification in Russia, the UK and Germany as opposed to, say China, Nigeria and France? There are two reasons for this. First, on an epistemological level, while all these countries are located in Europe, they have very different histories and state structures, resulting in different relationships between capital, state, land and the people. Despite morphological similarities, they present different sets of institutions, relations and practices. As the study focuses on causality, not regularity or generalisability, choosing other cases would have allowed including even more dissimilar cases, making it possible to find other causalities. However, this would hardly have resulted in substantial changes with regard to the general theoretical argument put forward. As Jennifer Robinson has argued, comparison can start from 'anywhere' (2016) and there is no privileged position of one potential case against the other. Second, there are practical issues. Analysing the relationship between gentrification and public policies on an in-depth basis in different countries demands the researcher have the capacity to read local literature, analyse planning documents and legislations, and conduct interviews.[5] Furthermore, background knowledge on history, culture and social structures is helpful, to say the least. As German, Russian and English are the only three languages I speak to a degree that allows for this kind of work and as I didn't have the resources to hire a team of researchers, the selection of cases has been influenced by my personal background. I also have known all three cities for a long time, which made it easier to build upon existing contacts and obtain access to various sources of information.

With regard to data and sources, the research strategy has been different for all three cases studied.

In this respect, London is the birthplace of gentrification as a concept and arguably one of the top sites covered by urban studies worldwide. When trying to understand the constellations here, I was able to draw on an expansive and varied literature. Moreover, due to the concentration of academic institutions in London, I profited enormously from the background knowledge and advice provided by fellow researchers who have studied London for decades. I also interviewed a limited number of experts on the neighbourhood I researched, including real estate managers and planners. In addition, compared with Germany and Russia, the range and the quality of statistical data is excellent and the data was accessible.

My research on Prenzlauer Berg differs from this. Prenzlauer Berg is not only the place I have lived in for most of the last 30 years, but it is also a neighbourhood that I have studied intensively since I became an academic. As a consequence, most of the data and much of the interpretation included in this book are based on past work that I have brought together and updated for this study. Methods used here include a literature review, the analysis of public planning documents, interviews and participant observation in numerous events, meetings and public hearings over the last three decades.

St Petersburg has been a completely different affair. The term gentrification has hardly made it to Russia thus far, and the number of studies on this phenomenon is very small. Moreover, in Russia, the transition from socialism to capitalism has resulted in taking away resources from the academic system to such a degree that urban research is generally fairly weak (when compared with the West). The staggering lack of resources has also contributed to isolating many Russian scholars from discussions in the West and this has been changing only very recently. Moreover, official statistics are notoriously unreliable in Russia. In order to grasp the dynamics under which gentrification operates there, I therefore had to do primary research. In this, I was supported by Russian colleagues who were enthusiastic about the issue and willing to invest their time and intellectual resources. Together with them, I organised a Research Laboratory in 2015 at which we discussed the applicability of Western literature to the Russian situation and performed empirical research in two neighbourhoods destined for larger urban renewal projects in St Petersburg. We applied a mix of document analysis, expert talks and intensive site inspections accompanied by interviews with residents to obtain an understanding of the actual developments on the ground. For a number of reasons, this work was never finished – yet it provided invaluable insights that helped me to understand the specificities of the situation there.

I proceed as follows: having sketched my approach in the previous pages, Chapter 2 re-examines the rent gap. Its task is to recapitulate the central features of this theory, as well as to discuss its limitations. Doing so, I try to avoid 'the booby trap of the stalemate between "production" versus "consumption" explanations' (Slater 2017, p. 121) that has occupied gentrification research for a long time (for summaries of these debates see Lees et al. 2008; Brown-Saracino 2010). I find that the arguments brought forward in this debate have reached a point of saturation where conceptual progress becomes unlikely and I aim at new lines of questioning. The main message I develop in this chapter is that the rent gap theory is indispensable for explaining gentrification, but too abstract and oversimplifying. I discuss five major problems in this context that point to the limits of this theory. These are: a lack of attention towards barriers to capital flows; the *problematique* of connecting the scales at which potential and actual ground rents are produced; the implications of a nomothetic conceptualisation of land rent; the limitations of a unidirectional understanding of property relations as 'control'; and turning a blind eye towards the actual conditions for the realisation of rent increases and the role that the state plays in this. On this basis, Chapter 3 provides a first look into the institutional parameters that frame the operation of gentrification. It analyses the different housing systems in the UK, Germany and Russia through a historical perspective that highlights the strategies, ideologies and struggles that have shaped the design of the diverse institutions at different points of time and discusses the implications of these for the operation of the rent gap. Taking a long-term perspective, I focus on the changing conditions for reinvestment in the housing stock, roles of tenures and rent laws, and identify 12 commodification gaps (see Table 3.6) relevant for the operation of gentrification in the UK, Germany and Russia at different times. Together, these gaps guide the analysis of the cases on the neighbourhood level that follows and are further refined through empirical examination. Chapters 4–6 apply this framework to the neighbourhood level and examine the operation of gentrification in the neighbourhoods of Barnsbury (London), Prenzlauer Berg (Berlin) and the centre of St Petersburg (Russia). Finally, Chapter 7 discusses the implications of this analysis for the theorisation of gentrification. It revisits the main themes, discusses the findings of the empirical observations and suggests a reconceptualisation of gentrification. It refines the commodification gap as a new theoretical concept and argues for a reorientation of gentrification studies.

Notes

1. In a similar vein but with different language, David Ley has characterised the gentrification concept as 'promiscuous' and as having 'a never-ending appetite to include more and more areas into its study' (presentation at the RGS-IBG Annual International Conference 2016, 'Planetary Gentrification', Authors-Meet-Critics session, 31 August 2016, London).

2. While the debate following Jennifer Robinson's book *Ordinary Cities* (2006) has become too extensive to be repeated here, three accusations can be summarised that form the core of many postcolonial critiques (see McFarlane 2010). First, it has been argued that urban studies have been dominated by Western parochialism and it has been suggested that theories and methodologies that were developed in the West were too easily applied in settings 'foreign' to their place of origin. Second, there is criticism that urban studies have been marked by a tendency to study the 'usual suspects'; without including more dissimilar contexts, the opportunity to learn from differences is severely diminished, resulting in dominating paradigms being reinforced. Third, it has been pointed out that much of the debate has been marked by a search for 'superlative', 'archetypal' and 'paradigmatic' cities (see Beauregard 2003; Brenner 2003), leaving little space for cities that would not fall into this realm.

3. I acknowledge that demand-side explanations have helped a great deal in understanding gentrification, and I also accept that supply will never be delivered without demand, so understanding the demand for inner city housing and the underlying social, demographic and cultural patterns is important. Nevertheless, I refrain from including this wing of explanations in my study. This is for two reasons. The first is that gentrification has changed since the 1970s and it has become emancipated from the 'back to the city movements' experienced in the USA and the UK at that moment in time. Theories developed against this background have, therefore, lost much of their former explanatory power. Providing evidence that gentrification in today's Mumbai, Moscow or Mexico City differ from those in Islington or the Lower East Side 30 years ago hardly comes as a surprise and, as a consequence, I find the benefit of such an undertaking questionable. This is also supported by studies from the Global South, most of which show only weak interest in the Western imagination of gentrification as an influx of middle-class professionals, artists or hipsters and their usual consumption patterns.

4. I should emphasise that this treatment of the term excludes a number of developments that appear in the literature but, in my view, should not be counted as gentrification. Thus, an 'incumbent upgrading' of the social structure resulting from an improvement of the social status of sitting residents should not be counted as gentrification because it hardly involves displacement. Neither should publicly financed improvements be regarded as gentrification, as long as they are not connected to market mechanisms and undertaken for the sake of a return on the capital invested. State-led displacement motivated by reasons other than profitability (e.g. ethnic cleansing) also stands outside this definition.

5. Researchers from the UK or the USA clearly profit from colonial history in this regard, as they can easily do comparative studies using their native language. Unfortunately, this also supports an imbalance within urban studies in which the UK and former British colonies are extremely overrepresented both as empirical cases and with regard to their impact on wider theorisations.

CHAPTER 2

Why the Rent Gap isn't Enough

The rent-gap thesis formulated by Marxist geographer Neil Smith in 1979 is one of the best-known, if not *the* best-known theoretical arguments about gentrification. The core idea is that gentrification is a structural product of capitalist land and housing markets. This conceptual architecture gives the rent-gap theory a decisively universalist flavour and makes it a top candidate for discussing the relationship of particularity and universality in gentrification research.

In the following, I will first describe the core arguments of this conceptualisation and discuss its main advantages in comparison with other explanations, before turning to the specific weaknesses of this theory and examining where the rent gap falls short and needs to be modified, opened and expanded.

Where the Rent Gap Works Well

The concept of a rent gap first saw the light of the day in an article entitled 'Toward a theory of gentrification. A back to the city movement by capital, not people' (Smith 1979). In the four decades that have passed since then, the rent-gap theory has arguably become the most well-known in gentrification research. While the father of this argument 'did not guess that anyone would take that paper too seriously' when it came out (Smith 2010, p. 97), it is now taught in undergraduate courses around the world. It has been subject to heated attacks and quarrelsome discussions (Ley 1986; Badcock 1990; Bourassa 1993; Hamnett 1991, 1992), stimulated the observation of other gaps (Hamnett and Randolph 1984; Sýkora 1993; López-Morales 2011)

The Commodification Gap: Gentrification and Public Policy in London, Berlin and St Petersburg,
First Edition. Matthias Bernt.
© 2022 John Wiley & Sons Ltd. Published 2022 by John Wiley & Sons Ltd.

and inspired troublesome efforts to operationalise and test it empirically (Ley 1986; Clark 1987, 1988; Kary 1988; Badcock 1989; Engels 1994; Hammel 1999a, 1999b; Yung and King 1998; O'Sullivan 2002; Diappi and Bolchi 2008; Porter 2012).

The essentials of the rent-gap argument are easy to explain. Its quintessence is already clearly expressed in the subtitle of the journal article in which the theory was first articulated (Smith 1979), i.e. 'A back to the city movement by capital, not people'. With this, Smith clearly distanced his approach from previous explanations of gentrification. Whereas these, Smith argued, had focused on the sociodemographic profiles of 'pioneers' and 'gentrifiers' moving into gentrifying neighbourhoods, a 'broader theory . . . must take the role of producers as well as consumers into account' (Smith 1979, p. 540). The major challenge for gentrification theory would therefore not be explaining 'consumer preferences', but studying why capital is directed into gentrifying areas. The answer provided in the paper is the rent gap.

For Smith, the emergence of a rent gap was embedded in the dynamics of residential property values in capitalist economies. In order to make his point, he engaged arguments from both classical ground rent theories and Marxian political economy. Four categories are important:

1. *House value*: Following a labour theory of value (Marx, Ricardo), the value of a house is defined as the amount of socially necessary labour power needed to produce the house. It is not to be confused with the price of the house (which is an outcome of the interplay of supply and demand), yet it sets the level from which the price fluctuates. Due to capital depreciation, the house value is likely to decrease in the long run, if no new labour is undertaken for maintenance, replacement, extensions, etc.

2. *Sales price*: 'Since the land is generally sold along with the structures it accommodates' (Smith 1979, p. 542f.), the sales price represents the value of the house, plus an additional component for the ground rent.

3. *Ground rent and capitalised ground rent*: 'Ground rent is a claim made by landowners on users of their land. . . . Capitalized ground rent is thus the actual quantity of ground rent that is appropriated by the landowner, given the present land use' (Smith 1979, p. 543).

4. *Potential ground rent*: Potential ground rent is the amount of ground rent that could be capitalised under the land's 'highest and best use' (a term Smith borrowed from von Thünen, a classical theorist on land rent).

Smith related these categories to the life cycle of residential properties in inner city neighbourhoods (see Figure 2.1).

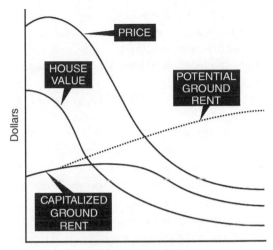

FIGURE 2.1 The evolution of the rent gap (Smith 1996, p. 65).

In short, the argument depicted by the diagram in Figure 2.1 works as follows. When a building is newly constructed, its house value, its sales price and its capitalised ground rent are high. During the first cycle of use, the ground rent and sales price are even likely to increase as urban development continues outwards and, as a consequence, the centrality of the property (hence the rental payment it can achieve) is on the up. Over time, however, this situation is likely to change. Without further upgrades or considerable maintenance work put into the building, its value is bound to decrease and, together with it, the capitalised ground rent it can achieve. As a consequence, the sales price falls too, making under-maintenance and 'milking the property' a rational option for the owner to still achieve a satisfying revenue flow. If this is done by the majority of landlords in an area, the outcome will be a speeding-up of house value, ground rent and sales price deterioration, finally pushing the actually capitalised ground rent under the level of the potential ground rent (i.e. what could be gained for the property under the highest and best use. Thus, a rent gap emerges, i.e. a 'disparity between the potential ground rent level and the actual ground rent capitalised under the present land use' (Smith 1979, p. 545). For Smith, this rent gap is essential to the possibility of gentrification, since it is only when this gap is wide enough that gentrification will be initiated and capital will flow into the housing stock:

> To summarize the theory, gentrification is a structural product of the land and housing markets. Capital flows where the rate of return is

highest, and the movement of capital to the suburbs along with the continual depreciation of inner-city capital, eventually produces the rent-gap. When this gap grows sufficiently large, rehabilitation (or for that matter, renewal) can begin to challenge the rates of return available elsewhere, and capital flows back.

(Smith 1979, p. 546)

With this argument, Smith took a radical departure from both neoclassical economics and the Chicago School of sociology, which were mainstream amongst urban scholars in the 1970s (for a detailed discussion, see Lees et al. 2008, chapters 2 and 3) and demonstrated how the workings of markets together with social inequality resulted in transforming spaces to the highest and best use for the better-off, while creatively destroying shelters needed by less advantaged social groups. What makes the theory special is, as Slater put it, 'its critical edge, its normative thrust' (Slater 2017, p. 119).

In its initial formulation, the rent-gap theory is based on two theoretical cornerstones that are brought together and applied to a decisively urban situation. It is essential to discuss the rent-gap theory 'in tandem' (Slater 2017, p. 124) with these broader conceptualisations in which it is situated.

The first of these is the theory of uneven development (Harvey 1973, 1982, 1985; Smith 1982, 1984). Already in Neil Smith's 1979 piece, references to the theory of uneven development are easily visible, although in a rather infant form. Here, Smith still reluctantly argued that the 'process is cyclical and, because of the long life and fixity of such investments, new cycles of investment are often associated with crises and switches of the location of accumulation' (Smith 1979, p. 547). Three years later, in an article on 'Gentrification and uneven development', he made the link to the rent- gap theory explicit:

The logic behind uneven development is that the development of one area creates barriers to further development, thus leading to underdevelopment, and that underdevelopment of that area creates opportunities for a new phase of development. Geographically, this leads to the possibility of what we might call a 'locational seesaw': the successive development, underdevelopment, and redevelopment of given areas as capital jumps from one place to another, then back again, both creating and destroying its own opportunities for development.

(Smith 1982, p. 151)

The parallels with the rent-gap argument are palpable: capital invested into the built environment gets 'frozen' for a while, thus, effectively leading to disinvestment and abandonment, and, by the same token, creating new opportunities for investment. Abandonment and reinvestment are, thus, linked, and the decline of an area is an opportunity rather than a barrier for its upgrading.

The second theoretical cornerstone of the rent-gap thesis, are Marxist theories of ground (or land) rent. Originating from the works of classical economists like Smith, Ricardo, Malthus and Mill, these theories define land rent as the sum of money paid by the user to the owner of a parcel of land. The main difficulties in discussing land rent arise from the fact that land is a very special commodity: land can be utilised for the purpose of production without being produced itself, its supply is naturally limited and inelastic, every piece of land is unique, its value is only realised over long periods and it includes long-term relationships between owners and users. Reacting to these idiosyncrasies, Karl Marx (most of Smith's treatment of the issues is based upon Marx's work) put a great deal of work into elaborating 'the shitty rent business' (Marx, cited by Ward and Aalbers 2016, p. 1760). In the course of this work, Marx developed different definitions of land rent. Building on earlier work from David Ricardo, he first argued that different qualities of land (referring to agriculture, he actually spoke of fertilities) and the enhancements invested in it, justify differential rents. As the conditions for the emergence of different qualities of land can be caused by natural features or by man-made improvements, he named these 'differential rent 1' and 'differential rent 2'. Second, Marx held that every rent is a 'monopoly rent': as land is scarce and every piece of it unique, its owner holds a monopoly over the land and prices are only limited by effective demand. Third, and this is what mainly distinguishes Marxist theory from other approaches, Marx argued that rent is a social relationship between the landlord and the tenant that is built on property ownership. Hereby, rent is the form in which this relationship gets realised economically.

It is easy to see how this conception lies at the foundation of Smith's rent-gap theory: plots of land in cities have different locational qualities, thus achieving different prices. The competitive advantages of one piece of land against another can change, e.g. in the course of suburbanisation, so that the ground rent demanded for it will depreciate. As a consequence, differential rent paid for this piece of land will suffer and be pushed below the potential ground rent. At the same time, investments in the land can lead to higher productivity (Marx's differential rent 2), thus closing the rent gap and making sure that the highest and best (and most profitable use) is realised. As this change affects the relations between landlords and tenants, it is not just a neutral economic factor, but translates into the displacement of one class of tenants by another.

Summing up, it can be said that the rent-gap theory can be seen as a combination of ground rent with the theory of uneven development. Its main strength lies in the development of 'a synthetic conceptual tool that has been a consistent application of rent theory at the urban level' (Ward and Aalbers 2016, p. 1776). On this basis, three advantages can be identified that make the rent gap an indispensable tool for any theorising about gentrification.

First, the rent-gap argument provides an explanation of why and how the capacities for gentrification (i.e. renovated houses, infrastructures, shops) are provided. It, therefore, does not consider individual consumer preferences irrelevant (despite some taunts on 'consumer sovereignty explanations' by Neil Smith; see for example Smith 1996, pp. 49f.), but focuses attention on the 'decision-making environment' (Hamnett and Randoph 1984) in which agents involved in producing gentrification operate.

Second, drawing heavily on theories of uneven development, it fosters a historic view that enables one to interpret current rounds of upgrading as a consequence of previous neglect. Thus, the rent-gap theory promotes a perspective that goes beyond snapshot studies of gentrification and forces the researcher to take temporality and historicity, i.e. historical actuality, into account.

Third, the rent gap links economic conditions to social change. Attempts to increase the actual ground rent are linked to a change of use structures that can only be achieved by displacing low-income households and attracting the better-off. Contrasting notions of upgrading, revitalisation, or reorganisation, the rent gap thus includes social conflict and class structures in the picture.

Where the Rent Gap Falls Short

As with every theory, the rent gap has its limitations. Most of these are connected to the two theories (and to some degree even the specific variant of these theories) upon which it is based. As a consequence, criticising the rent-gap theory demands going beyond the narrow formulation of the rent gap in Neil Smith's writings and including its theoretical roots. Based on this, in the following I will focus on five interrelated issues that I regard as major limits of the rent-gap theory. These are (i) a lack of attention towards barriers to capital flows, (ii) the *problematique* of connecting the scales at which potential and actual ground rents are produced, (iii) the implications of a nomothetic conceptualisation of land rent, (iv) the limitations of a one-directional understanding of property relations as 'control', and (v) turning a blind eye to the actual conditions for the realisation of rent increases and the role the state plays. Thereby, my goal is not so much to 'falsify' the rent-gap theory; given the 'ingenious simplicity' (Smith 1996, p. 42) of the argument and its intentionally limited claims, this would be a pointless undertaking. What I intend to do instead is, rather, to find openings that take the rent gap as a starting point, but allow for the advancement of the argument conceptually. This, I claim, also leads to better empirical research, as it allows for the identification of gaps and shortcomings that have not yet been addressed systematically.

Before stepping into the ring with the rent gap, some words of caution are necessary. The first is that Marxist concepts such as the rent gap are always meant to describe structures on which social relationships are based rather than empirical phenomena. As the working of these relationships is time and place specific, there is no specific appearance assigned to this essence. The concept fluctuates and resists any attempts to tie it to a singular form.

Related to this, and this is the second precaution which needs to be held in mind, there is an important difference between the rigorously structuralist architecture of the rent-gap argument and the more open and political treatment of the issue in empirical studies. Most notably, this is also the case within the works of Neil Smith. Already in his undergraduate thesis entitled 'The return from the suburbs and the structuring of urban space. State involvement in Society Hill, Pennsylvania (Smith 1977), Smith devoted whole chapters to discussing the role of the state in underwriting, regulating and safeguarding the actions of private developers. This theme was frequently repeated in later works, e.g. on the 'revanchist city' (Smith 1996), and on 'waves of gentrification' (Hackworth and Smith 2001). In all these works, Smith undertook a fine-grained analysis of political shifts facilitated by the state to decipher how gentrification was influenced by the changing forms of interaction between private and state actors.

Beyond Smith's work, researchers have also developed alterations of the rent-gap argument that follow the basic route outlined by Smith, but are geared towards specific institutional and political environments. The most important of these are the value gap thesis by Hamnett and Randolph (1984) describing the 'flat break-up market' in London in the 1970s; the invention of a 'functional gap' emerging with the transition from a planned economy to a market system in the former Eastern Bloc (Sýkora 1993); the introduction of a scalar perspective by Daniel Hammel (1999a, 1999b); and the concept of two ground rents (a lower one capitalised by current owner-occupiers and a higher one capitalised by the market agents of renewal), respectively, i.e. two rent gaps, introduced by López-Morales for the case of Santiago de Chile (López-Morales 2011). What all these studies impressively demonstrate is that despite numerous critiques and serious difficulties with operationalising the rent-gap hypothesis and especially with exporting it to other contexts, Smith's explanatory approach has been a productive source of inspiration on the basis of which a fine-grained understanding of political economy underlying gentrification can be achieved in very different contexts. While this can be regarded as a proof of the strength of the overall argument, the multiplication of gaps also leaves us with an unpleasant juxtaposition of the rent gap as a universal explanatory device and the need to employ a number of other, more particularistic gaps when this general theory is applied to different situations.

This, however, reflects a problem: On the one hand, the rent-gap argument is based on a theoretical framework that treats gentrification as a tendency endemic to all capitalist cities, notwithstanding their particular historic, cultural and political constellations. On the other hand, empirical studies of gentrification either need to depart from this argument and treat it as a mere background for rather historical and political lines of reasoning, or modify the rent gap to make it fit with different institutional contexts. Whereas in the first perspective the rent-gap argument and the politico-institutional environment it is embedded in are analytically separated, in the second perspective, the argument is tailored to fit to the particular environment in which the study is located, and it could almost be argued that every case justifies its own concept of a gap. In the following, I will argue that this situation reflects five limits of the rent-gap theory and its failure to take the institutional contexts into systematic account.

When and Why does Capital Flow?

The first of the five limits has to do with the discussion of capital flows. Here, the basic idea put forward by Smith (1996, p. 65) is that underinvestment (or the growth of the potential ground rent) opens up the rent g and once the rent gap 'grows sufficiently large . . . capital flows back' (Smith 1979, p. 546) and renewal is set in place. There are several issues with this that are explored below.

First, while the functioning of this mechanism can easily be demonstrated in the cases where it is set into motion, observers have long found it troublesome to work with the concept in places where the rent gap is wide, but capital still doesn't flow back in. The problem here is that the decision to invest capital into a piece of land is not only dependent on the conditions of the piece of land and its surroundings, but on the wider social and economic environment in which the investment decision is being made. While the conditions for the opening of the rent gap are produced locally, the likelihood of reinvestment is also determined regionally, nationally and even globally.

Second, 'capital switches' do not happen automatically. Quite in contrast, there are numerous barriers to the flow of capital through which the movement of capital is restrained. Building on Marx, Harvey (2010) has discussed at length six potential barriers to capital accumulation in general that can lead to a 'switching crisis':

i) insufficient initial money capital; ii) scarcities of, or political difficulties with, labour supply; iii) inadequate means of production, including so-called 'natural limits'; iv) inappropriate technologies and organisational forms; v) resistance or inefficiencies in the labour process; and vi) lack of demand backed by money to pay in the market.

(p. 47)

All six barriers are familiar to urban scholars. The lack of capital and labour are regular companions of both big projects and more small-scale undertakings. Especially in underdeveloped economies, the difficulties of finding credit has proven to be a well-known obstacle to many investment plans. Technological and natural limits cause a great deal of concern for engineers and planners, and it is not always possible to overcome these barriers, e.g. when a development is to occur on swampy land or in a flood-prone area. Investment plans are regularly met with concerns by the users already occupying the land affected (environmentalists, preservationists and other groups), and the outcome is protests that are capable of making the investment more expensive or even stopping the investment altogether. Finally, the history of urban development is full of examples of speculations that have failed, and in many cases developments have fallen vacant or a climate of low demand has caused demotivation.

In sum, although a piece of land might be underdeveloped and ripe for investment, there are manifold reasons why this investment might not occur and blockage at any one of the points I have discussed can disrupt the continuity of capital flow (Harvey 2010). The exploitation of a rent gap is, thus, not automatic at all, and the availability of capital and the attractiveness of investing it into land vis-à-vis other opportunities is anything but self-evident.

This has also been acknowledged by Neil Smith who, in earlier works (1977), empirically demonstrated how capital needed to be underwritten, lured and directed by state subsidies to allow for reinvestment. Since these early works, state complicity has been a recurrent theme in critical gentrification research (Smith 1996; Hackworth and Smith 2001; Allen 2008; Hodkinson 2012; Kallin and Slater 2014) and how the state has worked as an active facilitator of gentrification has repeatedly been studied. The major theoretical implication of all these studies is to acknowledge that the flow back of capital is only a hypothetical tendency and that there is a need for extra-economic forces to help overcome potential barriers to the movement of capital.

The problem, however, is that much of this literature could be read as if states would always and automatically work in the interests of capital – while empirical evidence shows that this is only rarely the case. In reality, the role of the state is much more varied and multifaceted and, in addition to fostering gentrification, there are also instances where capital switches have been made difficult by state actions or where states have simply been incapable of supporting investors. Thus, exclusively focusing on the state as a supporter of gentrification leads to a one-sided and incomplete picture.

At Which Scale is the Rent Gap Positioned?

The second concern is the spatial scales at which actual and potential ground rents are determined and the question of how these are interlinked.

The first person to address this *problematique* was Daniel Hammel, in a paper called 'Re-establishing the rent gap: An alternative view of capitalised land rent' (Hammel 1999a). He argued that:

> Potential land rent is determined at the metropolitan scale, that is, by the factors that work at the scale of an entire city. . . . Capitalised land rent is determined largely at the neighbourhood scale. The general socio-economic characteristics of the neighbourhood, including land use, act to limit land rent. Thus, the capitalised land rent of a particular site may be less than its potential if the land use of the surrounding parcels is not of the type that will allow the full measure of potential land rent to be captured.
>
> (Hammel 1999a, p. 1289)

In other words, while an area might potentially be ripe for gentrification, there can be factors that limit its realisation, so that capital doesn't flow in and the rent gap remains. Hammel tailored his argument at a neighbourhood scale and showed how a number of factors at the neighbourhood level (e.g. a bad reputation of an area, particular ownership structures, neighbourhood associations, etc.) could counteract the tendency of capitalised ground rent to fall and, thus, create disinvestment or operate in a way that would make reinvestment difficult.

Yet, this line of reasoning can be taken further up and down the scalar hierarchy.

Thus, it can be argued that the capitalised ground rent is not constituted at the scale of a neighbourhood, but at each individual parcel. As every piece of land is unique, it can be thought of as forming its own monopolistic market, as a man-made island of absolute space (Harvey 1974, p. 249). There are as many land markets as there are land parcels and property owners. Conceptually, this wouldn't be a problem if all owners acted according to neoclassical assumptions, i.e. if they would all behave rationally, share a preference for maximising the return they can gain from their property, had the same access to the means necessary for doing so (knowledge, capital, management capacities) and had equal access to adequate information that would allow them to understand the exact point at which the rent gap is wide enough for their property to make reinvestment attractive. In reality, this is hardly ever the case. On the contrary, information is scarce, capacities are unequally distributed, and even economic rationality as an exclusive motive is questionable.[1] What follows is that valorisation strategies can be different from parcel to parcel. Whereas one owner might decide to disinvest and get his return from saving on maintenance costs, his neighbour to the left might fix her property up and achieve higher rental incomes, while the neighbour to the right could be a charity organisation and provide sound accommodation at below market rates for altruistic reasons. In all three cases, the level of the capitalised ground rent would differ. To continue with this example, the first owner might act as 'rational' profit maximiser and sell out once the rent gap is wide, the second might prefer to hold her property and increase rents, while the third might find himself uninterested in taking action. Again, the strategies for operating the rent gap would differ significantly. Thus, as much as every piece of land forms its own market, every property can be said to have its own rent gap and, hence, the question arises of how these individual gaps can be aggregated to bigger units to form effects at the neighbourhood level. Necessarily, this leads to questions about the scale at which upgrading and displacement needs to be observed to qualify as gentrification.

Structuralist scholars have usually found it hard to accept the relevance of individual agency with regard to landlords and claim that there would be a tendency towards the equalisation of landowners' strategies as an outcome of globalisation and financialisation (see, for example, Harvey 1982), so that the actual agents of the rent gap can be abstracted from. Yet even when it is accepted that the existence of rent gaps is an outcome of the wider economic environment and cannot be made dependent on the capacities and intentions of individual property owners, the issue isn't solved but becomes even more complicated. The reason for this is that it is impossible to define the actual value of a potential ground rent. How can we know what the highest and best use for a parcel is? Back in the 1970s, the answer to this question seemed more or less obvious (at least for residential markets): a higher and better use could be achieved when more or better-off users were attracted. At this time, property markets were, however, still local in nature and rent levels were determined locally by surplus, production differential, and the rent-paying ability determined inside city boundaries. It is debatable whether this picture still holds true. Nowadays, as a consequence of globalisation and financialisation (Aalbers 2012, 2016; Gotham 2009; Lapavitsas 2013; Rolnik 2013), previously separated land, housing and financial markets have become integrated and investment decisions are not necessarily guided by considerations of local conditions alone anymore. In the real estate sector, land and housing markets have become increasingly merged with financial and investment markets and different land markets (e.g. housing and office, or the markets in New York and Berlin) have begun to operate in much closer proximity to each other. As a consequence, capital easily switches between different forms of land use. The more housing investment has become globalised, the more the expectations of investors reference global benchmarks. In addition, investment funds and other financial instruments nowadays usually merge fairly diverse 'enhancements' (e.g. condo flats in Canada, shopping malls in Slovakia, and gated suburban communities in Spain) under one investment scheme, pack them into bundles with other assets and buy and sell them on the stock markets. All these changes have made real estate very liquid and brought housing closer to financial products.

The implications of financialisation for gentrification theory are considerable. Calbet i Elias has pointed out four ways in which financialisation can push up the potential ground rents in a specific location (Calbet i Elias 2018, p. 168):

1. The internationalisation of real estate markets may trigger price prospects that are not based on local demand and price levels, but on housing markets in other (sometimes disparate) contexts and geographies.
2. The predominance of investors with high access to capital results in greater price competition and higher price costs.
3. The predominance of institutional investors with high return expectations has the consequence of price prospects that are orientated towards the exchange value of not solely real estate but also financial investments.
4. The increasing prices tend to inherently produce speculative spirals.

Under these conditions, rent is not merely a payment from a land user to its owner, but it has become intertwined with financial speculation. As a consequence, it has become next to impossible to find out if an investment is taken for the sake of achieving the highest and best use for a particular piece of land, as collateral against other investments or for other reasons. The potential ground rent of an actual piece of land has, thus, become increasingly decontextualised from the local context (Savini and Aalbers 2016). The emergence of 'planetary rent gaps' (Slater 2017) in which different spatial scales are interwoven and potential ground rents are increasingly defined through globalised expectations, however, entails serious consequences for the rent-gap theory. If the potentiality for higher rents is not determined by more affluent users, but by opaque capital flows on the planetary scale, the link between rent, locality and better-off users, which is central to the rent-gap argument, is considerably blurred and it becomes questionable how potential ground rents, reinvestment and a change of resident structures can still be bound together.

Which Rent?

In a similar vein, the treatment of rent invites questions too. As already noted, the concept of rent is one of the two theoretical pillars of the rent-gap argument. It is all the more surprising that the treatment of rent, as such, is rather simple.

Re-examining the urban rent discussion blossoming in the 1970s, it becomes quickly clear that talking about rent can refer to very different social phenomena. Marx distinguished between absolute rent, differential rent 1 and differential rent 2, and monopoly rent. There has been considerable debate about the essence and the limits of all these conceptualisations among Marxists. Furthermore, over time, even more concepts of rent have been suggested, including class-monopoly rent (Harvey 1974), redistributive rent (Walker 1974) and fiscal rent, global rent and derivative rent (Haila 2016). All these different concepts express different dimensions, constellations and agencies related to the extraction of ground rent.

Regrettably, Smith's formulations of the rent gap and most theoretical contributions building on it have been fairly abstinent in regard to making use of these discussions. In almost all cases, when a term like potential ground rent is used, we hardly find a hint of how rent is defined. It is, thus, not clear whether the rent gap is about absolute rent, differential rent 1 or differential rent 2, monopoly rent, or if rent is taken as a pure metaphor for studying historically specific relationships between landowners, users and state actors in a rather empirical manner. The problem here is that – as Table 2.1 sketches – different treatments of rent suggest different causations for the emergence of rent gaps and point towards different social relationships.

Thus, if a rent gap were to be understood with regard to absolute rent, it would be based on speculative landholding and hoarding. If it were seen as an expression of differential rent 1, the reason for its emergence would likely to be found in changing locational patterns, which, for example, would bring middle-class residents in closer proximity to new employment opportunities. If it were conceptualised in relation to differential rent 2, the focus would be on different housing qualities, enabled through renovation or densification. If the focus were on monopoly rent or class-monopoly rent, one would focus on the ability of landlords to create submarkets and defend them from

TABLE 2.1 Classical forms of rent and gentrification

Forms of rent	Description	Modern examples with relation to gentrification
Absolute rent	Rent arising because of the existence of a class of landlords acting as a barrier to entry for capital or consumers.	Speculation on prices increases by property owners.
Differential rent 1	Rent arising from increases in productivity due to some feature of the land.	Rents demanded for close proximity to job or leisure opportunities or other infrastructures.
Differential rent 2	Rent arising from increases in productivity as a result of investment in the land.	Rents demanded for the higher quality of housing after renovation.
Monopoly rent	Rent arising from some unique, non-substitutable feature of the commodity.	Rents demanded with regard to the unique and exclusive character of a neighbourhood (e.g. ethnic homogeneity, beachside location).

unwanted uses. In short, as plausible as the use of the category rent for the explanation of gentrification may appear at first glance, with a more detailed look it becomes obvious that different concepts of rent point towards very different social relations.

Examining the deficits of the rent-gap theory at this point, one should have in mind that the rent-gap theory was formulated in the midst of a revival of interest in classical rent theories in the 1970s. However, as early as the 1980s, studies on land rent started to be divided into different camps, and this led to considerable differences in the treatment of the issue (for a history of these debates see Haila 1990; Ball 1985b; Jäger 2003; Park 2014; Ward and Aalbers 2016). Within this context, the rent-gap argument is clearly positioned within what Haila (1990) termed the 'nomothetic' argument led by David Harvey.

What are the main characteristics of a nomothetic understanding of rent? What did these entail for the formulation of the rent-gap theory and why can this be seen as problematic?

A major characteristic of the nomothetic camp of rent theorists can be found in the 'search for a general theory, which flies in the face of the acknowledged diversity and heterogeneity of landed property relations' (Haila 1990, p. 283). While much of the research of that time had moved from simple conceptualisations of landowners as a homogeneous class to an interest in studying the diversity of property relations, Harvey (1982) moved in the opposite direction. He acknowledged heterogeneity in principle, but postulated a tendency towards uniform behaviour, resulting from a tendency towards an increasing treatment of land as a financial asset. With this, Harvey distanced himself, to some degree, from focusing on rent as a social relation (something which he had emphasised a few years earlier; see the section 'Property as Control?') and looked at the role of rent as a coordinating device for the circulation of capital. The issue was, thus, shifted from studying rent as a societal phenomenon, rooted in social relations between users and owners, to focusing on investment flows into the built environment and the coordinating role rent played in these.

Unsurprisingly, this raised many critiques. Behaviouralist studies took great effort to identify different types of agents involved in the development of land markets, e.g. landowners, developers and finance providers. The outcomes were typologies and studies on the behavioural characteristics of different sorts of property owners, mostly with an emphasis on those that are distinct from those implied by the assumption of rational profit maximising actors (see Massey and Catalano 1978; Goodchild and Munton 1985; Kivell 1993). In summary, these studies showed that the motives of landowners were fairly varied and that a tendency towards equalisation was not supported by empirical evidence. Building on this literature, Healey and Barrett (1990) argued that Harvey and his followers merely managed to establish a tenuous relation between the structures of land rent dynamics and the specific interests and strategies of individual agents acting on this terrain.

The tension between grand theories and actual empirical evidence was even more fundamentally criticised by Michael Ball, who demanded to look at 'detailed historical situations rather than to make gestures towards some grand general theory' (Ball 1985a, p. 86). He developed a 'structures of provision' approach that focuses on describing 'contemporary network(s) of relationships associated with the provision of particular types of building at specific points in time' (Ball 1998, p. 1513). For Ball, the structures for realising rent were of a historically contingent nature and what were demanded, therefore, were 'concrete analyses of which take account of social relations and the historical development of the area in question' (Ball 1977, p. 402), instead of generalisations. On this basis, Ball went as far as to suggest that urban land rent theory would be outlived:

> In one respect, the approach suggests that urban rent theory is dead. The search for categories of rent leads to an emphasis on an urban land market whose existence as a universal entity across all urban space is

> highly questionable, and whose existence at specific, limited, and histori-
> cally changing locations within any urban area is not independent of the
> process of creating the built environment itself. In other respects, how-
> ever, the argument suggests that rent theory does have an important role
> to play: when limited to the independent landowner and the impact on
> urban land use of the appropriation of that social agent's revenue as rent.
>
> (Ball 1985b, p. 523)

Whether Ball really argued in favour of dispensing with the concept of rent altogether or just wanted it to be used in a more specific way, remains unclear. At some point, the debate arrived at a stalemate and nomothetic scholars could not be convinced to give up their category fundamentalism (Ball 1987, p. 269), nor could idiographic researchers take the step from empirically describing historically specific structures of division to developing a new general theory. As a consequence, the interest in theorising land rent considerably cooled down.

Summarising the debates of this time, it nevertheless becomes clear that the concept of rent should not be applied in a purely abstract way, but rather as a means to address historically specific constellations. In other words, what was established is that 'in order for land rent to be a useful tool, it should be theorised in the context of its institutional embedding' (Jäger 2003, p. 245). Up until now, this call has, unfortunately, not found much reverberation in gentrification studies. More often than not, rent is treated as a self-evident category without going into depth about the specific relations enabling it.

Property as Control?

A similar problem can be observed when discussing the role of property within the development of the rent-gap literature. Following Marx, the per-spective applied here is largely an individualistic one[2] that sees rent (hence rent gaps) 'aris[ing] because there exists a class of owners of resource units – the land and the relatively permanent improvements incorporated in it – who are willing to release the units under their command only if they receive a positive return above some arbitrary level . . .' (Harvey 1974, p. 241). Con-ceptually, the idea that rent arises from the control over a piece of land has been at the core of land rent theories since the seventeenth century. Land rent, to quote Marx, presumes 'that certain persons enjoy the monopoly of disposing particular portions of the globe as exclusive spheres of their private will to the exclusion of others' (Marx 1971 [1894], p. 752). In a way, the concept of a monopoly control is fundamental to the concept of land rent, as it determines the reason for the emergence of rent as payment from users to owners. Without conceptualising a group of people holding some sort of control over access to a piece of land, there would hardly be a way to conceptualise land rent as a payment based on a social relation between the owners and the users of land and the division lines to the neoclassical treatments of rent (as a payment made for factors of production that are inelastic in their supply) would get blurred.[3]

While the idea of a monopoly control over land is central to the theory of rent and, hence, the rent gap for conceptual reasons, it is also difficult to handle and problematic when applied in the real world. Two problems stand out and are now explored.

I start with rent as a social relation. Here, Harvey and Smith clearly distanced themselves from the neoclassical treatment of rent as a production factor and set their work in the tradition of classical rent theorists like Petty, Smith, Ricardo and (most notably) Marx, who all saw rent as primarily a transfer payment to the owners of a particular piece of land. The point here is that this transfer payment is not just a technical operation, but includes a social relationship between the owner of a particular parcel of land and its user. Writing tongue in cheek, Harvey pointed out that 'actual payments are made to real live people and not to pieces of land. Tenants are not easily convinced that the rent collector merely represents a scarce factor of production' (Harvey 1973, p. 251). In the text that followed he gave numerous examples of the dynamic character of this relationship. He elaborated about struggles between landlords and low-income tenants, as well as about conflicts between speculators and middle-class suburbanites, and made clear that the realisation of land rent depends upon the ability of one class interest group to exercise its power over another class interest group. As a consequence, if and how land rent can be realised is a matter of class power and conflict from this perspective, and what tenants and other users can win in this struggle is what landowners lose. In the words of Harvey: 'The appropriation of rent, in short, entails the exploitation of who by whom?' (Harvey 1982, p. 333).

Somewhat paradoxically, there is also a second, quite different, reading of land rent in Harvey's and Smith's papers. Here, both repeatedly and obstinately use the term monopoly control. In Western societies, Harvey wrote, 'rent is, in effect, a transfer payment realised through the monopoly power over land and resources conferred by the institution of private property' (Harvey 1974, p. 240). This was nearly literally quoted by Neil Smith who argued that 'ground rent' (hence rent gap) could only be gained because 'private property rights confer on the owner near monopoly control over land and improvements, monopoly control over the uses to which a certain space is put' (Smith 1979, p. 541). In a footnote on this, Smith went even further: 'Certainly zoning, eminent domain, and other state regulations put significant limits on the landowner's control of land, but in North America and Western Europe, these limitations are little more than cosmetic' (Smith 1979, p. 547, fn. 7). While Harvey and Smith, thus, acknowledged the existence of limits to landowners' control in principle, they found them negligible and not worthy of being integrated into their theoretical considerations.

Moreover, it is not exactly clear which form of social relations Smith and Harvey had in mind when using the term monopoly. One way of reading the argument might refer to the everyday use of the term in economics. Here, a monopoly is defined as the market control that exists when a specific person or enterprise is the only supplier of a particular commodity, so that he or she can dictate prices. For land, this is obviously not the case as there is a multiplicity of property owners. Thus, a monopoly could only be claimed if it is assumed that all different owners of land shared the same goals and acted like a cartel in order to be able to demand monopoly prices. The key problem, then, is the aggregation of different individuals under the banner of a common class interest (to always achieve a positive rate of return) and the equalisation of this interest with collaborative action. Here, the argument implicitly assumes that there are identical interests between different owners and, moreover, that these either automatically or by some unspecified form of coordination, lead to common action. Even if one buys into the idea that different landowners have identical interests (which is empirically wrong), it is hard to imagine how they would facilitate a common strategy without assuming that they would either have a 'natural' sense of always knowing what their best interests are or appealing to a rather conspiratorial idea of coordination taking place in some staunchly committed secret organisation.

On an empirical level, the concept of monopoly control, thus, already invites questions. There is also, however, a second, deeper, and more theoretical issue with the concept.

Legal geographers (Blomley 2003, 2004, 2005; Graham 2011) and legal realists (Heller 1999; Singer 2000; Merrill and Smith 2001; Katz 2008; Penner 2009) have long criticised the reduction of the complex social and political dimensions of property to a unidirectional act of domination as simplistic and individualistic, and have highlighted an understanding of property in land as necessarily relational. As the right to control, change and exploit things does entail powers over others, these authors have argued that ownership has a dyadic, or even triadic, character: it influences the life chances of the non-owners of property and it is only made possible by the fact that others, either directly or by means of a legal system, recognise ownership rights. In this sense, property is not static and pre-given, but it depends on a continual, active 'doing' (Rose 1994), on a continuous enactment and re-enactment (Blomley 2004) of property rights. Property is, thus, about 'regulating relations among people by distributing powers' (Singer 2000). This relational character is only poorly understood with a conceptualisation of property as control.

While this is not the place to dive deeply into the different conceptualisations of property in the philosophies of law, legal geography or property rights theories in economics, it is worth pointing to the concept of ownership as a 'bundle of rights' that is used by many researchers to better grasp complexities of property ownership (see Honoré 1961; Patterson 1996). The basic idea of the bundles of rights metaphor is that property entails a bundle of social relations that are expressed in a bundle of different rights held not only by the owner, but also by third parties, the public and the government.[4] The most commonly known modern formulation is offered by Honoré (1961, p. 113), who listed 11 incidents of ownership: the right to possess, the right to use, the right to manage, the right to the income of a thing, the right to capital, the right to security, the right of transmissibility, the right of absence of term, the duty to prevent harm, liability to execution, and the incident of residuarity. Ownership, in this conception, is not regarded as a thing or as a single social relation between owner and user, but as a bundle of codified social relations in which certain interests are protected against others. The concept of owning a house, to give an example, would need to be substituted by considerations about the different rights connected to the house. The owner might, thus, have the right to use and occupy, mortgage, lease, sell or subdivide the property, yet at the same time the government might prohibit noxious or nuisance uses, make the property subject to historic preservation, collect taxes on the property, enforce a lien, impose zoning or even declare eminent domain. Furthermore, third parties might have use rights as tenants, further restricting exclusive 'near monopoly control'.

In sum, property should be seen as a bundle of relations, rather than conceptualised in terms of control and domination. The way this issue is approached by the rent-gap literature, thus, obfuscates rather than clarifies very complex systems of rights and duties.

How is the Rent Gap Realised?

A fifth, problem of the rent-gap theory lies in the question of how reinvestment and displacement are conceptually linked. This issue is not explicitly dealt with in the rent-gap theory, yet it is of crucial importance, as social class change is the essence of gentrification.

Interestingly, in this field, Smith seemed to see a working of simple market mechanisms as self-evident, with no need for further discussion. It seems that Smith regarded the issue in a circular way: as capital is only invested when higher returns can be achieved, a closure of the rent gap must be accompanied by increases in the price of housing, which in turn excludes users with inferior buying power, so that gentrification is achieved. The way displacement is thought to work here has nicely been described by David Harvey through the analogy of filling up places in a theatre:

> In the housing market with a fixed housing stock the process is analogous to filling up seats sequentially in an empty theatre. The first who enters has n choices, the second has n-1, and so on, with the last having no choice. If those who enter do so in order of their bidding power then those with money have more choices, while the poorest take whatever is left after everyone else has exercised choice.
>
> (Harvey 1973, p. 168)

The point seems very simple: when labour and money are put into a building, and it is upgraded and modernised, its price will increase. As a consequence of unequal income distribution, the effects of price increases are felt differently by different social groups. As capital is only invested when a satisfactory rate of return can be achieved and as the rent paid by the residents is the only source of return within the field of housing, putting capital into the existing building stock is only attractive when prices go up. This results in poor households with low spending capacities being excluded and replaced by better-off residents. Thus, in a market economy, unequal income distribution, competitive bidding power and the opportunity to demand higher rents for upgraded housing stocks work together.

How useful, though, is such a view? Is it a description or an explanation? The point here is that in reality, the price for housing is hardly ever only an outcome of a competitive bidding process in which well-to-do people can afford to outbid all other groups for land, but it is also defined by the regulatory frameworks in which this bidding takes place. Far from being just a matter of spending power, the displacement of low-income residents is co-determined politically, and the ways in which land use, house prices and rents are regulated have far-reaching consequences for the dynamics of urban upgrading

The state–market nexus in the provision of housing has been at the core of comparative housing studies for decades (for an overview, see Doling 1997; Lowe 2011; Arbaci 2019). Blossoming in the 1990s, this strand of research conceptualises housing as a part of welfare provision and analysed systemic differences in the construction of housing markets. The starting point of these studies is that housing is neither completely provided by the market nor completely decommodified in most capitalist societies. It is at the same time a commodity, produced for the purpose of capital accumulation, and a social right regulated by the state. As a consequence, housing can be adequately studied with neither purely economic nor with purely political-historical conceptualisations.

One of the most influential arguments on the state–market nexus in the provision of housing has been made by the Swedish sociologist Jim Kemeny. Very much influenced by Esping-Andersen's take on the *Three Worlds of Welfare Capitalism* (1990), Kemeny identified two essentially different housing regimes operating in Western countries (Kemeny 1995, 2006; Kemeny and Lowe 1998). Thus, while more liberal welfare regimes like the USA, the UK, and Australia were characterised by a dualist rental market in which ownership is the dominant form of tenure and the state only takes responsibility for the provision of rental housing for the lowest strata of the population, other states have institutionalised unitary rental markets or integrated markets. Here, social and private renting are integrated into a single rental market and housing subsidies are provided to all types of housing providers. In such a market, one can find a fairly wide range of different types of landlords (municipal companies, housing associations, cooperatives, private landlords) that all operate in the private rental market. Closely related, tenants enjoy more legal protections, and rents, contracts and standards are more influenced, if not controlled, by the state. Owner-occupation is, therefore, less desirable and makes up a significantly smaller share of the overall housing stock.[5]

What is the relevance of this literature for the "realisation problem?" Clearly, if one follows the argument that the price of housing is not only determined by supply and demand, but also by forms of state regulation and intervention, the explanatory value of approaches that stem from this perspective on the interrelation of capital flows, ground rent dispossession and differentiated buying power is limited. The concept of housing systems, thus, provides a necessary complement.

In this sense, Table 2.2 demonstrates how different forms of decommodification impact on the closing of the rent gap by building a barrier to the full realisation of the potential ground rent, by taking segments of the housing stock out of the market or by providing subsidies so that the revenues for the highest and best use are at least covered by the state, not the resident.

Thus, if parts of the housing stock are managed as social housing, rents are determined by administrative procedures, not the interplay of supply and demand. The construction of social rented housing has for a long time been one of the most severe and, at the same time, most visible, forms of state intervention into housing markets in many countries. With great differences in ownership structures, rent levels, funding streams, private sector involvement and rent setting (see Balchin 1996; Whitehead and Scanlon 2007; Scanlon and Whitehead 2008), the one thing that is common to this enormous

TABLE 2.2 Decommodification and displacement

Form of decommodification	Relevance for gentrification
Social housing	Rents are subject to political determination and administrative procedures in the stock affected and are not an outcome of supply and demand.
Homeownership	(Outright) Homeowners hold comparably strong rights that shelter them from market pressures. At the same time, a home owned is an asset that can be mortgaged, sold, and rented out and can be used for gaining a profit on the market.
Rent regulations	The price of housing is restricted by regulations, so that it does not follow market logic alone. However, segments of the market and types of residents and landlords can be affected in very different ways.
Allowances	Additional income enables low-income households to live in an acceptable standard of accommodation at a price they can afford.

variety is that social housing was, or is, built with the strong support of the state. Social housing stands to some degree outside the market and even if there is a potentially higher ground rent, it cannot be realised as long as the status as social housing is in effect. Moreover, with a large share of social housing, the level of market prices is depressed, so that the potential ground rent is reduced.

Subsidising homeownership has a more ambivalent effect. On the one hand, at a legal level, homeownership provides the owner with certain rights that in most capitalist societies are more comprehensively designed than user rights. Being an outright homeowner equals increased autonomy and shelters the affected household to some degree from market pressures. In effect, it is exactly this comparatively strong legal status of property ownership that makes it the preferred form of tenure in many countries. On the other hand, there is also an economic dimension to homeownership: a house or a flat that is owned is an asset.[6] As houses can be exchanged, consumed and circulated in markets, they become a means with the help of which wealth can be accumulated, stored and transferred between individuals and across generations. Once bought, houses constitute assets that can be used to buffer a household against unforeseen needs and on which equity can be drawn. At the same time, buying a home is normally the biggest investment taken by a household in the course of a lifetime, and for most households it is only possible with the help of a mortgage. This links households to financial markets, and as the subprime crisis in 2008 demonstrated, makes them vulnerable to developments occurring in these markets. Subsidising homeownership reduces this dependency to some degree by lowering the costs of credit. The relationship between individual homeownership and gentrification is, thus, contradictory. On the one hand, homeowners are sheltered against displacement pressure and can enjoy stable housing costs when their mortgage is paid off or when the payments are subsidised to some degree. On the other hand, the higher the potential ground rent increases, the stronger the incentives to monetise the asset.

Rent regulations, as they are in many integrated markets, cap the price of housing at a pre-determined level, thus decreasing the level of the potential ground rent for the unit affected. Rent regulations, however, differ highly between countries. Lind (2001) distinguished two dimensions along which rent regulations are designed:

> The first dimension concerns whether the control covers rent changes for sitting tenants or rents generally. The second dimension is whether the aim is to protect the tenants against rents over the market level, against sudden big increases in rents or if the aim is to keep rents permanently below market levels in attractive areas.
>
> (Lind 2001, p. 41)

While rent regulations, in general, slowdown rent increases and buffer the affected stocks against the market, the actual effect depends very much on the particular framework. The whole issue becomes fairly muddled when looking at the details, and the outcomes usually are complex geographical patchworks between rents following market logic and those determined politically.

Thirdly, housing allowances have an impact too. In contrast to 'brick and mortar' subsidies, which aim to reduce the price of a dwelling, income related housing allowances focus on the demand side of the market. In general, the development of housing allowances as a policy tool reflects a shift in the role of the government in the housing market (Kemp 1997; Turner and Elsinga 2005). It has enabled governments to move away from expensive supply-side subsidies, which were central to the housing policies of many countries in the decades following World War II and move towards a form of support that was seen as a more efficient because it was more means tested. Housing allowances are a common feature of advanced welfare states that enable low-income households to live in an acceptable standard of accommodation at affordable prices. By subsidising the buying power of a household, allowances build a buffer against displacement, without depressing the market price. They, thus, assist in achieving and even increasing the potential ground rent, yet they pass the costs of this, to some degree, to the taxpayer, not the individual resident. The effect of this is a weakening of the link between rent increases and displacement.

Summarising, capitalised ground rents are not only determined by supply, demand and differentiated bidding powers, but also by political determinations. So there is a state–market nexus in the production, consumption and valorisation of housing that has been intensively discussed in the field of housing studies. Integrating this research into the explanation of gentrification can, therefore, arguably produce a more complete and less simplified understanding of the phenomena and complement the economic explanations discussed.

Embedding Gentrification

Throughout the earlier part of this chapter, I have discussed the theoretical contours of the rent-gap argument and carved out some of its conceptual limits. My main criticism towards the rent-gap theory is, in a nutshell, that it suffers from an oversimplified perspective on the way markets work. It can, therefore, make a convincing general argument about the general economy of gentrification, but is of very limited use for understanding its political preconditions.

What becomes clear from this discussion is that gentrification cannot be isolated from the context in which it takes place. While it rests on the working of financial, real estate and housing markets, these markets are also socially produced. The point is not to show that capital flows in one or the other direction and that this can lead to pricing out low-income residents in one way or another, but to understand the conditions under which this is made possible. What needs to be explained then is when, where and how capital enters into the production of housing, how this is guided by different actors with different interests and capacities, how their actions are shaped differently in different contexts, how the investment of capital impacts on housing rights held by different groups, what the effects on the price of housing are and why and how this results in displacement.

Economy, Society and States

Economy, society and states must not be separated in the analysis of urban upgrading. Instead, they should be interwoven and integrated into one concept that allows us to consider the operation of the economic and political conditions for gentrification in tandem.

However, how can the relations between markets, states and societies be adequately conceptualised? When answering this question, it is helpful to take a look at conceptual debates in the field of economic sociology that have developed a particularly fine-tuned understanding of the interrelations of markets with states and societies (Polanyi 1957 [1944]; Granovetter 1985; Hall 1986; Esping-Andersen 1990; Fligstein 1990, 2001; Evans 1996; Callon 1998; Crouch and Streeck 1997; Hall and Soskice 2001; Smelser and Swedberg 2005; Swedberg 2004). A second stream of theory building that is important for this study is to be found in state theories, namely, in what has been termed an integrated conceptualisation of the state (Gramsci 1971; Poulantzas 2000 [1978]; Jessop 1990). In the following, I summarise the major propositions of these two debates, before moving on to the question of how they can inform research on gentrification.

Central to economic sociology is the notion of embeddedness. It goes back to Polanyi's famous book *The Great Transformation* (1957 [1944]), but has really only come to prominence with Granovetter's work (1985), which insists that all economic activities depend on networks of social relations, not only pre-capitalist activities, as Polanyi argued. The point here is that market societies – even the most 'liberal' ones – are intertwined with non-economic institutions. Consequentially, markets and societies cannot be separated but are intermingled and need to be studied together. In this view, the economy is regarded as an interconnected web of organisations interlinked with state institutions and society at large. In the field of housing, for example, mortgage banks, developers, land use planners, regulatory bodies, the construction industry, tenant organisations and governments are all part of a system of housing provision (see Ball 1998) in which they interact in interrelated and dynamic ways. In the words of Fligstein and Calder (2015, pp. 1–2):

> Markets are socially constructed arenas where repeated exchanges occur between buyers and sellers under a set of formal rules and informal understandings of governing relations among competitors, suppliers, and customers. These rules and understandings guide interactions, facilitate trade, define what products are produced, sometimes constitute the products themselves, and provide stability for buyers, sellers, and producers. In modern capitalism, markets also depend on governments, laws, and larger cultural understandings.

States and markets are, thus, 'joined at the hip' (Fligstein and Calder 2015, p. 4). At the same time, the relationship between them is anything but stable and monolithic. Quite in contrast, political-economy literature on comparative capitalisms (Esping-Andersen 1990; Hall and Soskice 2001) has impressively documented that:

> . . . the relationships among governments, workers, and capitalists have varied dramatically over time and geography, and that economic trajectories are often culturally and nationally specific. Markets are social and political constructions reflecting a country's culture, its history of class relations, and the various interventions its governments have carried out through history.
>
> (Fligstein and Dioun 2015, p. 71)

While the state is, thus, crucial for the working of markets, it is itself far from being an autonomous organisational entity. This is not the place to dive deeply into the waters of state theory – but it can be held that today an understanding of the state as monolithic, independent and autonomous from society has largely been overcome. Developing a balance between recognising the relative autonomy of the state as an institution that is both separate from the market and civil society, while at the same time acknowledging its intermingling with these, has been a core theme of state theories throughout the last century and has occupied generations of theorists. Here, a line of enquiry has become most useful that was pioneered by Antonio Gramsci (1971), advanced by Nicos Poulantzas (2000 [1978]) and brought to widespread recognition by Bob Jessop (1990, 2002, 2015). It is often referred to as the 'integral theory of the state'. In this understanding, the state is its own institutional form, with specific organisational structures, procedures and logics – but its very architecture, strategies and instruments reflect the contradictions of the society and are open to political contestation. In the words of Nicos Poulantzas (2000 [1978], pp. 128–129), the state:

> . . . must be considered as a relation, more exactly, as a material condensation (apparatus) of a relation of force between classes and fractions of classes as they are expressed in a specific manner (the relative separation of the state and the economy giving way to the very institutions of the capitalist state) at the very heart of the state.

In this view, states are regarded as sets of apparatuses (but also as systems of rules and procedures) whose sections and levels serve as power centres for different social groups or fractions of social groups (see also Jessop 1990). They are, thus, more of a network of organisations, rather than an organisation in its own right. In understanding the work of this assemblage, two concepts introduced by Bob Jessop (1990) are useful. The first is the notion of a state project. It is closely connected to the Gramscian notion of hegemony and describes the possibility of competing strategies and hegemonic projects within the state. As Jessop argues:

There is never a point when the state is finally built within a given territory and thereafter operates, so to speak, on automatic pilot according to its own definite, fixed and inevitable laws. Nor, to be somewhat less demanding, is there ever a moment when a single state project becomes so hegemonic that all state managers will simply follow universal rules to define their duties and interests as members of a distinct governing class. Whether, how and to what extent one can talk in definite terms about the state actually depends on the contingent and provisional outcome of struggles to realise more or less specific 'state projects'.

(Jessop 1990, p. 9)

These reflections suggest that state strategies, instruments and actions in general need to be understood not as the unified expression of a dominating logic or principle (as functionalist theories of the state would argue) or as an expression of the common will of the dominating class carried out by 'special bodies of people' (as Lenin held), but as 'emergent, unintended and complex resultant of what rival "states within the state" have done and are doing on a complex terrain' (Jessop 1990, p. 9).

While state form and state action are themselves an expression of balances of power, historic blocs and hegemonic projects, the actions of the state nevertheless have a differentiated impact on the capacity of various social groups and classes to pursue their interests. State policies necessarily 'offer unequal chances to different forces within and outside that state to act for different political purposes' (Jessop 1990, p. 367); they privilege some interests over others, advance some goals over others and will be more accessible by some groups than by others. State institutions can, therefore, be used by economically powerful elites, and often are, but they can also form bridgeheads of resistance for the popular masses.

The Commodification Gap

Summing up a complex debate, it can be held that markets are intertwined with states and societies. Markets are embedded into societies, and economic actors must have an appropriate social setting to emerge from and be reproduced. As a consequence, economic transactions, states and political struggles need to be understood as intermingled, forming a dynamic field of relations. This field is marked by contradictory and constantly moving relations of force between different social groups that condensate into institutions. As these institutions selectively privilege some interests over others, they have a formative effect on the ways in which markets are built and the rights of some groups are advanced (e.g. property owners). Political interests are, thus, built into the very construction of markets. Thereby, it needs to be made clear that the terrain described must not be misunderstood as a simple binary that puts deregulation (the unleashing of market forces) on one side and regulation (protection against the market) on the other. Instead, both the ways in which markets are constructed and the design of protections against markets are subject to state action, and political interests are built into the architecture of the economy in multiple and intricate ways. To make things even more complicated, regulations can be the outcome of nationally, regionally and locally specific compromises that have condensated (Poulantzas 2000 [1978]) in dissimilar ways and materialised in numerous institutions at different spatial scales. Contradictions and tensions between different state projects dominating the regulatory landscape in different sectors (e.g. mortgage regulation vs. welfare) and spatial scales (e.g. national vs. local state agencies) are, therefore, the norm rather than the exception.

How can these conceptualisations inform the conceptualisation of gentrification?

Based upon the ideas sketched above, the directions are summarised as follows. First, and most importantly, economic processes can only meaningfully be studied when they are set in their political context. Instead of explaining gentrification on the basis of generalised rules of capital accumulation and land rent capture in the first place and bringing in the actual historic configurations under which these are made possible as context and variations as a second step, economic structures, and political struggles and state institutions need to be brought together from the start of the analysis. Second, the way this can be done needs to be historically and locally specific. Third, while historical analysis must give way to the inclusion of a broad variety of contingent factors, the findings need to be brought together in a concept that interweaves political and economic determinants of gentrification, instead of singling out one at the cost of the others.

Instead of analytically separating political struggles and the economy, this book, therefore, develops the concept of a commodification gap. I argue that focusing on the nexus between commodification and decommodification in different historically specific settings enables us to overcome the reductionist theoretical architecture of the rent-gap argument and put political struggles at the centre of explanations.

I define the commodification gap as the disparity between the potential ground rent level that can be achieved for a piece of land when it is fully commodified and the actual ground rent capitalised when it is decommodified, partly decommodified or non-commodified. I argue that it is only when this gap is closed that investment in housing becomes a viable option and gentrification is set into action. Capital flows where the chances to realise a maximum rate of return are highest – but these chances are defined by the state, as an outcome of social conflict and political contestation. It is only when the barriers to an uplifting of the capitalised ground rent are taken away and when land is predominantly a commodity that reinvestment can compete with the rates of return available elsewhere. This allows capital to flow into housing, thus, enabling its gentrification. Gentrification, in other words, demands the relative weakness, inexistence, bypassing or lifting of components of housing provisions that are decommodified. When this is not achieved, gentrification doesn't take place.

The crucial point made here is that the way in which commodification and decommodification work is historically and geographically specific. Decommodification can be achieved in different forms; it is built into uneven geographic constellations and produced by essentially various interplays of markets, states and societies. State projects, capital flows and ground rent work together to form commodification gaps. It is for this reason that the general theoretical concept of the commodification gap needs to be specified and applied in close relation to the context in which it emerges. This gives way to a number of individual commodification gaps that work in a singular context, but not for others. Table 3.6 presents examples of these gaps based on the empirical research presented in this book.

While details on the substance of these gaps will be discussed later in this book, what they all have in common is that they represent the intermingling of politics, institutions (like social housing or rent regulation) and rent gaps.

Commodification gaps can only operate in tandem with rent gaps since they can only emerge when there is a difference between the actual ground rent and the potential ground rent for a piece of land. With such a gap, the capitalised ground rent is depressed and prevented from achieving its highest and best use and reduced to a lower level. The rent gap and commodification gap, therefore, need to be read together, not separately.

Gentrification is, in other words, caused by general conditions – the existence of land rent, uneven development, and unequal income distribution – that apply everywhere there is capitalism and particular conditions that only work at a specific time and place. The rent gap and commodification gap are, thus, interconnected.

How far does this conceptualisation differ from the theories discussed earlier in this chapter?

First, in contrast to many postcolonial accounts, I don't agree that contextual differences make the rent gap useless. While the specificities of Neil Smith's (1979) analysis were indeed tied to a context specific to the USA at that time, the general argument was much broader and it is hard to imagine how, in a capitalist economy, any investment in housing would be done without a chance of profiting from increases in the capitalised ground rent. For this very reason, the rent gap is indispensable when analysing urban change within a capitalist economy. What is needed, therefore, is not 'lay[ing] the concept to bed' (Ghertner 2015), but more research on how the rent gap operates in different institutional environments.

Second, I agree with Lees et al. (2016) that the concept of gentrification has global application, but I doubt that applying enough 'flexibility to accommodate changing conditions and local circumstances' (Lees et al. 2016, p. 203) will suffice for understanding the actual ways in which it proceeds locally. If research is carried out more flexibly, but without an adequate theoretical apparatus that allows us to understand what observable differences between cases stand for, more and more single-case studies are produced – but the cases can hardly be brought into conversation on a broader level.

Finally, I also diverge from the work of Gavin Shatkin (2017) discussed in Chapter 1. By decontextualising the rent-gap theory and paring it down to the minimum, Shatkin treats the rent gap and its institutional environment as separate. This is echoed by many contributions on the Global South that describe gentrification in terms of a frontier (Gillespie 2020) between market and non-market forms of land allocation. Contrary to this, I aim to integrate political and economic factors into a single concept. I turn Shatkin's approach upside down and start with the institutions allowing for decommodification to explain the chance for the rent gap to be put to work.

By developing and applying the concept of a commodification gap, this book primarily aims to advance the theoretical debate around gentrification. It expands the concept of the rent gap by including the institutional conditions for its operation into the very core of the argument and integrates new ways of seeing the relationship between universality and particularity when doing urban research. My argument, thus, simultaneously builds on the rent-gap thesis and departs from it by not treating institutions as external, but placing the political embeddedness of markets into the centre of the explanation.

Notes

1. Massey and Catalano (1978), for example, have studied different types of landowners in England (large aristocratic estates, the Church, banks, insurance companies, pension funds, individual owners, corporate owners and government agencies) and have shown how all these different groups have distinct interests and modes of behaviour. Similar results have been found in studies in

Germany, where striking differences between 'valorisation-oriented' (*verwertungsorientierten*) and 'non-valorisation-oriented' (*nicht-verwertungsorientierten*) property owners have been observed in a rapidly gentrifying area (Reimann 2000).

2. It deserves to be noted, though, that this understanding is most pronounced in the early works of Harvey. In later writings, the economic functions of rent have been emphasised more intensively. Nevertheless, the understanding of rent as a social relation based on monopoly control has never completely disappeared.

3. Moreover, the idea of a social class holding exclusive control over land is not only theoretically coherent, but also in line with the historical development of Western societies in the eighteenth and nineteenth centuries. Be it British gentry, Prussian Junker, or Russian Boyary, a landed aristocracy was clearly distinguishable as a social group at that time.

4. There are differences, however, in the formal treatment of property between Roman Law (which is applied in most of continental Europe) and the Anglo-Saxon Common Law tradition. Whereas Roman Law only allows absolute ownership while recognising rights of non-owners, under Anglo-Saxon Common Law, ownership can be divided among different title holders

5. It should be noted that Kemeny's typology is anything but uncontested. In fact, there are a number of other typologies with slightly different conceptual backgrounds. Thus, Donnison, provided a 'logics of industrialism' perspective to the field (Donnison 1967) and argued that the expansion of state intervention into housing was closely tied to the overall economic development of a country. Barlow and Duncan (1994) developed a classification with a closer connection to Esping-Andersen's model and suggested a fourth ('rudimentary') cluster in which there was no strong state intervention into housing, and self-help, family support and a welfare role of the church remained important. Harvey, then, applied a Marshallian framework of social rights to the field of housing and argued that the real difference between different European Union states was seen in the degree to which housing was a constitutionally guaranteed right (Harvey 1994). Thus, housing policies cannot only be clustered into different families, but also sorted along a continuum ranging from weak/no rights to direct or indirect rights. More recently, Schwartz and Seabrooke (2008) suggested a 'variety of residential capitalism' approach and discussed the influence of the liberalisation of financial markets on housing systems.

6. This makes homeownership a cornerstone of individual asset-based welfare strategies and brings it in close relationship to the general welfare system of a society. For Kemeny (1995, 2005), there is even a trade-off between homeownership rates, tax policies and welfare spending: the financial burden for home buyers is quite extensive. Kemeny argued that they would be

predisposed to support low-tax policies that in turn would reduce welfare state capacities and lead to a low-welfare state. A related argument has also been made by Castles (1998, 2005), who argued that by retirement, owners would usually have paid off their mortgages and are then left with a net benefit equivalent to the rent they would otherwise have had to pay: 'In other words, when individuals buy their homes, they can get by on smaller pensions' (Castles 1998, p. 18).

CHAPTER 3

Three Countries, Three Housing Systems

The following chapter shifts the focus from the theoretical critique to the empirical analysis to examine how reinvestment and displacement have been intermingled with state action in three different countries and cities over time. I explain how different institutional configurations have resulted in the formation, opening and closing of different commodification gaps, and how this has enabled or restricted gentrification in a variety of nationally and historically different ways. In this chapter, the focus lies at the national scale. I examine how historically specific conditions for gentrification have been formed, before analysing the ways they have operated at a neighbourhood scale in the next chapter.

The British Experience

For any outsider, the two most striking characteristics of the British housing system are the conjunction of tenure and sociospatial segregation and the complex ways in which the relationship between the two has changed over time. These changes are so essential to the functioning of the whole system of housing production and housing consumption in Britain that a historian described the shape of the British housing system as being an outcome of two 'tenurial revolutions' (Daunton 1987) with prolonged aftershocks. Each of the two revolutions changed the existing configurations of housing fundamentally and led to very different dynamics between residents, classes, investors and the state.

The Commodification Gap: Gentrification and Public Policy in London, Berlin and St Petersburg,
First Edition. Matthias Bernt.
© 2022 John Wiley & Sons Ltd. Published 2022 by John Wiley & Sons Ltd.

From Private Landlordism to a Dual Market

The first of these revolutions emerged as a consequence of a number of developments that took the stage in the UK around World War I (see Merrett 1979; Merrett and Gray 1982; Daunton 1987; Holmans 1987; Ravetz 2001; Lowe 2004, 2011; Malpass 2005). Until that time, most of the housing in the UK was, as in much of Continental Europe, provided by private developers and built for renting. However, the existing form of housing provision was barely sufficient to deal with the challenges of massive urbanisation. In general, a dramatic shortage of affordable housing, widely spread overcrowding and a growth of slums were notorious in Victorian Britain (and dramatic in London). The conditions under which the urban poor had to live were in fact so problematic that even the aristocratic upper classes gradually understood that they were not an outcome of deviancy and lack of proper behaviour, but rather were affected by the labour and housing markets, and could only be improved by appropriate economic and social policies. The last two decades of the nineteenth century saw a broad movement of philanthropic experiments (see Bowie 2016), parliamentary debates and governmental initiatives that all searched for ways to deal with the inability of the existing system to provide adequate housing for the urban poor. Among the many initiatives of these times, the Housing of the Working Classes Act of 1890 played a particularly important role, as it broke with the liberal tradition of laissez-faire and first defined a context in which local authorities were allowed to intervene in the housing market and provide housing on their own. However, not much happened in the short term.

The late Victorian housing market got stuck in a model in which private rentals provided by a large number of small landlords with low rates of return and limited economies of scale formed the backbone of housing. Over time, the economic conditions of being a private landlord became even worse, and rental housing fell into structural trouble.

Yet, 'the decisive blow to the private landlord was dealt by the introduction of rent control in 1915' (Clarke and Ginsburg n.d., p. 7). The cessation of new construction during the war and the boom of war industries had resulted in a situation where the already existing shortage of housing was deepened. This stimulated ferocious rent increases in a number of cities, which were met by an outburst of resistance. Especially in Glasgow, where the militant Shop Stewards Movement was strongest, intensive rent strikes were set in place and the unions even threatened to go on general strike in the war industries until rent controls were introduced. In this situation, the government was quick to react and introduced the Rent and Mortgage Increase (War Restrictions) Act, which limited rents and mortgages to pre-war levels. Although this was originally planned only as a temporary measure, successive governments found that removing this type of protection for tenants would potentially be harmful to their electability. Thus, with the short interruption of a gradual rent decontrol implemented by the Conservatives between 1957 and 1965, it took until 1988 for the private rental market to return to the state of no-regulation experienced in Victorian times.

The aftermath of the war then resulted in a breakthrough that would completely change the structure of the British housing system for the rest of the twentieth century. Facing an even more imbalanced housing market with the return of demobilised troops, the Lloyd George government introduced a first national building programme in 1919, promising to deliver 500,000 'homes fit for heroes' (see Swenarton 1981) in three years. As it was clear that the private sector would simply not have the capacity to deliver on these ambitious goals (see Lowe 2004, pp. 166–167), the programme was initially entrusted to local authorities. This was the start of a long history in which the state came to play a dominant role in the provision of low-cost housing for the working classes, leading to a marginalisation of private land-lordism. Although the subsidies provided by the central government would originally not discriminate between public and private developers, successive programmes (most of all the Wheatley Act) would lay the basis for massive housing construction under the auspices of local councils. Consequentially, the 1920s saw a major expansion of the council housing sector: of the roughly 1.5 million apartments that were built between 1919 and 1930 in the UK, around one-third were built by public authorities. Labour-led councils, in particular, increasingly came to see the provision of council housing as a centrepiece of their municipal agenda and took pride in providing modern homes of high standards at low costs.[1] For many left-wing politicians, the expansion of council housing was even seen as a step towards a 'municipal socialism' in which the reproduction of the working classes would increasingly be organised through the local state.

Simultaneous to the expansion of council housing, the inter-war times saw an immense expansion of owner-occupation: 'Of the 4 million new houses built in Britain between the wars, over 70 percent were constructed by private enterprise, and . . . the majority were sold for owner occupation' (Malpass 2005). Three factors shaped this development. First, the availability of extensive public subsidies provided an incentive for construction. Second, simultaneous population growth and the restructuring of the economy worked together to develop a sustained demand from white-collar workers who were interested in respectable suburban housing and were capable of taking on a mortgage. Third, building societies became permanent institutions and increasingly took on the function of providing secure, low-interest mortgages that were central to financing the building boom. The outcome of these developments was a continued growth in homeownership: 'by the outbreak of the Second World War in 1939, nearly a third of households were owner-occupiers and the 'home-owning society' had been born' (Lowe 2011, p. 78).

Altogether, the inter-war years saw two fundamental ruptures with the system of housing provision that was in existence before. Both the emergence of council housing and the expansion of homeownership took place nearly simultaneously and resulted in what Ginsburg (1999) has called a 'liberal collectivist framework' characteristic for Britain's housing policy between 1915 and 1975. This framework had four defining features (Malpass 2005, p. 19):

• Rent control/regulation for private rented housing,
• Nationally subsidised provision of local authority rented housing (i.e. council housing),
• Slum clearance programmes with replacement by council housing, and
• Fiscal and general government support for owner-occupiers.

To summarise, these developments introduced a dual system of housing provision, in which owner-occupation and council housing would simultaneously become the norm and grow continuously.[2] Private renting, in contrast, was becoming a rather unattractive activity. While the system of rent controls set in place in 1915 was subsequently reformed in the 1950s–1970s to allow rents to be brought closer to market levels (mostly with conditions for new tenancies), there was no return to the laissez-faire environment of the Victorian times in this sector. The general pattern of marginalised private renting was in danger of extinction due to the flourishing of owner-occupation and a growing council sector kept intact over a long time.

The Thatcherite Revolution

All this changed when a Conservative government, with Margaret Thatcher as prime minister, came into power in 1979. The years to follow marked a decisive rupture with the inherited social compromise and changed Britain's housing system fundamentally. The overall policy approach of the Thatcher governments has often been described as a 'Conservative revolution'. It rested on three overarching aims (see Hodkinson and Robbins 2013, p. 59) that all marked a decisive break with the existing configurations of power. First, the – barely hidden – aim of many policies that were introduced (e.g. attacks on the trade unions, the abolition of the Greater London Council, cuts to public service delivery, etc.) was to weaken traditional strongholds of the Labour Party and ensure Conservative dominance in the political arena of

Britain. Second, a key ambition of the Thatcher government was to replace the existing 'welfarist' post-war consensus with a 'new conservative' common sense of individualism and entrepreneurialism, mixed with 'traditional values'. Third, against the background of a long post-World War II decline of the UK's economic power, a prime goal was to restructure the British economy along the lines of monetarist policies, strengthening competitive markets, price signals and consumer choice, and weakening all sorts of restrictions on the unchecked flow of capital and goods.

However, the 'roll-back' (Brenner and Theodore 2002) of established configurations was not put into action in a single policy, but rather through a number of mutually reinforcing initiatives. The flagship Conservative policy in the field of housing was without a doubt the Right to Buy (see Forrest and Murie 1988; Ginsburg 1999, 2005; Jones and Murie 2006), i.e. the introduction of a statutory right for sitting council tenants to buy their home, which was underpinned by enormous discounts (of up to 60% for houses and up to 70% for flats) off market prices. Until 2014, the Right to Buy resulted in the sale of 2.5 million council homes (all UK), making tenants themselves the main agents of privatisation. From the point of view of the Conservatives, introducing a right to buy was largely an end in itself. It was seen as a cornerstone for the development of a property owning democracy,[3] helping families to get out of houses owned by the state and fulfilling their dream of owning their own home. The core idea was that state provision of housing should be reduced as far as possible, enabling more citizens to show responsibility for themselves and their families by owning their house, thus, turning citizens from welfare recipients into asset holders. By and large, this policy turned out to be very popular – especially among those council tenants who had enough money to participate in privatisation and/or had the luck to sit in a well-located, high-quality or otherwise above average apartment. Non-white minorities also exercised their right to buy to an over-proportional degree, as a way out of the discrimination experienced in the rental market.

Sales of public sector dwellings rose dramatically in the early years of the Right to Buy policy and peaked in 1982, with 245,113 units (Jones and Murie 2006, p. 55). Altogehter, in the course of the 25 years after its introduction, more than 30% of council tenants exercised their right to buy. The outcome of this development was not only a considerable reduction in the public housing stock, but – even more importantly – an accelerated downgrading of the stock that remained. As right to buy sales were mostly concentrated in the most attractive estates, councils were disproportionately left with the problematic cases. Moreover, the balance between the people in need of social housing and the number of houses available for them changed considerably, making public housing more and more targeted towards those with the highest needs (e.g. the homeless, Black and minority ethnic groups, the unemployed, etc.) or with a long history of being on the waiting list. While council housing had provided shelter to close to one-third of all British households in 1981, this proportion decreased to 23% only 10 years later (https://data.london.gov.uk/dataset/housing-london). The outcome was an increasing 'residualisation' (Forrest and Murie 1983) of council housing, meaning that this type of housing came to overly concentrate low-income or economically inactive households, elderly people, single parents, ethnic minorities and other socially disadvantaged groups.

'The second, more insidious privatization mechanism was simply to turn off the tap of public investment in public housing . . .' (Hodkinson et al. 2013, p. 7). Through a combination of expenditure cuts and budget controls on local authorities, the Conservative central government minimised investment in the council housing stock, thus making privatisation more attractive for both tenants and local authorities. This was also facilitated by a new subsidy system that reduced central government grants to local authorities and implemented long-term increases in council housing rents to meet the resulting revenue gap. In a decisive attempt to reduce state expenditure, net capital expenditure into the housing stock was brought down to a level of about one-third compared to its peak in the mid-1970s (Malpass 2005, p. 106). If council housing already had an ambivalent reputation, it was certainly made worse under the Conservatives. From the tenants' point of view, 'the advantages of council housing ebbed away, as rents increased, maintenance and improvement withered, and "right to buy" sales visibly demonstrated government's lack of commitment to the sector' (Ginsburg 2005, p. 119).

The third major attack on council housing came with the introduction of 'voluntary stock transfers' through the 1988 Housing Act. In short, this policy promised to end the municipal monopoly on social housing and introduce a kind of choice for council housing tenants. It encouraged councils to enable the transfer of existing public housing to alternative and charitable landlords and gave tenants the statutory right to decide over the transfer of the stock they inhabited to an alternative owner. Although this policy started to be implemented very slowly, it gained speed in the 1990s and, as a consequence, from 1988 until 2008, 1.4 million dwellings were transferred across the UK (Pawson and Mullins 2010, p. 41). 'Britain's new social landlords' (Pawson and Mullins 2010) were mostly housing associations, i.e. private, non-profit organisations whose trading surplus is used to maintain existing housing and to help finance new homes. In contrast to councils, registered social landlords (RSLs), which is the official term for these organisations, would have more flexibility with regard to their projects and their staff, and the ability to raise money on capital markets and, thus, to fill the investment gap left by central government cuts. It was also argued that the new housing associations would be more professional than their precursors in general.

While large-scale stock transfers undoubtedly changed the ownership of much of Britain's housing, how this change is to be evaluated is still contested among academics and experts. Altogether, three main interpretations are on the table. The first suggests that the transfers were largely a defensive measure that allowed municipally managed housing to carry on in the face of a vigorous neoliberal attack. This view would point to the long-term partnership between councils and RSLs and emphasise the influence of local concerns on the RSLs' agenda. Empirical studies have, however, shown that considerable variety exists in the actual shape and intensity of partnerships between councils and RSLs and that the experiences seem to be fairly mixed. A second view interprets the changes as a form of modernisation of the public housing sector (Malpass and Victory 2010). Following a line of argument popular with New Labour politicians, this perspective argues that transfers created an opportunity for modernising the management, thus bringing social housing services closer to the residents and their needs, allowing more accountability and empowering tenants. Again, the evidence is mixed and empirical studies support both views. A third view (e.g. Ginsburg 2005; Watt 2009a; Hodkinson 2011a, 2011b; Hodkinson et al. 2013), which in the realm of politics is often identified with the Defend Council Housing campaign and its left-wing policies, has systematically criticised the transfers as some form of 'nicer' privatisation. Here, the line of argument focuses on the loss of long-term control, the exposure of housing to market risks (that comes together with more capital-based financing) and a weakening of staff and tenants' rights (DCH 2008). It is also pointed out that housing associations are increasingly run as private sector organisations, borrowing directly from banks, engaging in land speculation and 'for profit' housing developments and managing their stock as an asset to maximise returns. It is difficult to find a fair middle-ground between these opposing views, and the empirical evidence on this issue is mixed. Against this background, it seems reasonable to follow the account of Ginsburg (2005, p. 133) who suggested that 'social housing in Britain is gradually being commodified. It was, of course, a flawed service in many ways and there remains a "social" element in the new structures, but the notion of a universally accessible, comprehensive, affordable and electorally accountable rented housing as a public service is disappearing'.

At the same time as council housing was sold out and government support for this form of tenure squeezed, the Conservatives maintained support for owner-occupation through generous tax giveaways, like the exemption of revenues from capital gains and mortgage interest tax relief. As a consequence of the rapidly growing number of right to buy homeowners, the total volume of these tax reliefs quadrupled in the course of the 1980s (see Malpass 2005, p. 109), so that the costs of state support for this form of tenure were immensely expanded. The contrast with the cuts in the support for council housing is telling.

Moreover, in order to make capital for homeownership more easily accessible, the Thatcher government removed most of the boundaries between building societies and commercial banks, enabling building societies to demutualise, provide a broader range of financial services, change their status to a limited company and (most importantly) borrow on the financial markets. This lifted the fence between commercial banks and non-profit organisations in the financing of housing and brought mortgages in the sector more in line with those operating generally. 'This obviously increased the cost of a given loan, but it also brought supply and demand in closer relationship . . . Thus it became easier to borrow for house purchase, but more expensive' (Malpass 2005, p. 122). The demutualisation of building societies largely abandoned the existing pattern of long-term saving and low interest borrowing, and the whole market became more open and more dynamic. It linked housing finance closer with the general state of the economy, thereby fuelling the growth of homeownership, while at the same time exposing borrowers and lenders more fully to market forces.

In addition to supporting ownership – and running alongside the break-up of council housing – rent controls in the privately rented stock were completely abolished in 1988 (for details see Malpass 2005; Crook and Kemp 2011). Whereas previous reforms of rent regulations in the 1960s and 1970s had softened the system of rent control in place since 1915 but had not done away with it, the 1988 reforms completely deregulated this segment of the housing sector. In contrast to previous regulations, from early 1989 on, landlords were allowed to establish new tenancies on the basis of rents freely negotiated in the market. Moreover, 'assured shortholds' were introduced, which reduced the security of tenure to six months and enabled landlords to automatically get their property back after this time with a two-month eviction notice. Private renting was, consequently, made a very insecure form of tenure in which tenants would basically become exposed to the will of their landlords, with next to no protection. The aim of this 'Rachmanism[4] with tax breaks' (Ginsburg 1989, p. 60) was clearly to allow a higher return on capital and easier access to vacant possession, thus encouraging investment in private renting.

Integral to the new housing strategy, both for the private rental and the public sectors, was social tolerance towards higher rents. In the first year of operation of the new 1980 Housing Act alone, rents in council housing rose by an average 48% (Malpass 2005, p. 108). This, of course, also meant that more people were eligible for means-tested housing assistance. The Conservative government made a deliberate decision to shift housing subsidies away from bricks and mortar and towards housing assistance for individuals. The justification for this was mainly that individual means-tested benefits would be more efficient and targeted, as they would reduce the danger of spending tax money on supporting people who were not actually in need. Thus, not only was public intervention into the housing market downsized, but also its shape changed, away from delivering public housing as a social right and more towards a sort of welfare benefit for the 'deserving poor'.

Summarising the major developments of the 1980s, it can be stated that the British government at that time not only embarked on neoliberal rhetoric but also managed to change the existing configuration of housing in a fundamental way: it downsized the public housing stock and abolished protections for both public and private tenants, while at the same time expanding homeownership and making housing markets in general more driven by a laissez-faire environment. If the goal of the Conservative revolution had been to strengthen markets; restore a common sense of individualism, entrepreneurialism and consumerism; and lay out the foundations for lasting conservative hegemony, it had come a long way. However, as Hodkinson and Robbins (2013) have pointed out, the Conservative agenda was only halfway implemented, resulting in an 'unfinished revolution'.

New Labour: More of the Same?

More important were the developments in the years after 1997, when a New Labour government under Tony Blair came into power. While it was clear from the start that New Labour would not return to the policies of its precursors in the 1970s, it was nevertheless an open question as to how much it would be willing and able to roll back the Conservative revolution. The answer to this question, it soon became clear, was 'very little'. The cornerstones of Conservative housing policy were all maintained and the right to buy, the deregulations of rent controls in the private sector, and large-scale stock transfers were not abolished. Rather, 'Labour's approach towards council housing followed the Conservative's enthusiasm for privatisation and de-municipalisation, albeit under the guise of "modernisation". . . . Council housing was considered irredeemably "Old Labour" and hence unworthy of fiscal or political support' (Watt and Minton 2016, p. 210).

To understand why the Blair government acted this way, some more detailed remarks on New Labour's conceptions of the welfare state, in general, and housing policies, in particular, are necessary.

A much-hailed key concept of Blairite thinking at that time was the Third Way. Based on theories developed by Anthony Giddens (1998, 2000), spin-doctored by public relation officers (see Fairclough 2000) and globally popular with Social Democrats from the late 1990s onwards, Third Way approaches promised to reconcile right-wing and left-wing politics and advocated for synthesising individual liberty of strong market economies with social justice guaranteed by the state. Central to this kind of thinking was the idea that a modernised social democracy would not make either-or choices between the state and market, but rather would take the best from each. Typical for Third Way rhetoric were key terms that only provided vague direction and were often not very seriously conceptualised, but were taken together to provide a somewhat coherent narrative (Fairclough 2000). Central to the thinking employed by New Labour, in this respect, was a vocabulary of modernisation and pragmatism that would acknowledge that the world had changed, so that the 'old style interventions' of the Left as well as the laissez-faire attitude of the New Right would equally need to be transcended by a new thinking that envisaged politics between, above, and beyond established patterns; seeking more flexible, partnership-oriented and inclusive, rather than fundamentally statist or entrepreneurial way of thinking. While New Labour's philosophy has repeatedly been criticised (see amongst others Hall 1998; Levitas 1998; Fairclough 2000) and even reviewed as post-Thatcherism (Hall 1998), it nevertheless managed to provide a coherent hegemonic idea capable of implementing considerable change in British politics.

In this respect, three concepts that stood at the centre of housing and urban policy under the Blair governments deserve closer examination. The first of these is 'social exclusion'. The importance of exclusion visa-à-vis urban and housing policies was highlighted by the incoming Blair government as early as 1998, with the setup of a high-profile Social Exclusion Unit and the publication of a report titled 'Bringing Britain Together' (SEU 1998). Under New Labour, social exclusion became a central concern of government policies. The change from 'poverty' to 'exclusion' (with a strong emphasis on it meaning 'more than poverty') signalled more fundamental changes with regard to what were seen as the causes and the nature of social and urban problems. As Levitas (1998) has demonstrated, New Labour's understanding of exclusion conflated three distinctive and, in many ways, contradictory discourses: (i) a redistributionist discourse that emphasised poverty and material inequality (RED), (ii) a moral underclass discourse whose main concern was with morality and antisocial behaviour (MUD), and (iii) a social integrationist discourse that emphasised the significance of paid work for inclusion (SID). Levitas (1998, p. 27) wittily summarised: 'in RED they [the poor] have no money, in SID they have no work, in MUD they have no morals'. In broad terms, RED has always been favoured by the traditional left, whereas conservative and right-wing politicians have usually leant towards MUD explanations. New Labour's take on exclusion, in contrast, displayed all three explanations together in a flexible and changing way, and shifted from one to the other. In the policy arenas relevant for housing, however, concerns over morality and behaviour as well as over the significance of getting people into jobs tended to dominate New Labour thinking (see Levitas 1998; Watt 2009b). In this sense, New Labour's rhetoric worked as a justification for not addressing the problems of poor housing conditions and homelessness directly, and redefined those in terms of crime and antisocial behaviour (Watt and Jacobs 2000).

Rhetoric of 'local community' and 'active citizenship' has formed a second pillar of New Labour's urban political thinking (see Wallace 2016). Closely aligned to communitarian views (see, for example Etzioni 1996) and social capital concepts (Putnam 1993, 2000), a lack of bonds and a loss of community became central features in the diagnosis of urban problems. Consequently, empowerment and mobilisation of communities were increasingly seen as major preconditions for urban change. As with exclusion, the focus on 'community' raises a number of conceptual and empirical problems (Imrie and Raco 2003; Colomb 2007). In addition, Imrie and Raco (2003) have argued that putting the perceived breakdown of communities centre stage implicitly creates a subtle form of pathologising of the urban poor through a distinction between the deserving, competent, empowered and proactive citizens and the others (Imrie and Raco 2003). Moreover, a 'belief that empowered and mobilized communities can and should play an enhanced role in the development and implementation of urban policy agendas . . . requires a rearticulation of active citizenship, with the state's role moving from that of a provider of [welfare] services, to that of a facilitator – enabling communities and individuals to take more responsibility for the conduct of their own lives' (Imrie and Raco 2003, p. 235). An emphasis on community and state restructuring, thus, went hand in hand.

A third central concept in New Labour's urban thinking can be found in the application of the term social mix (see Colomb 2007; Lees 2008; Bridge et al. 2012). Again, the use of the concept is not particularly clear-cut and policy documents have frequently shuffled together functional, tenure and income mix. What is common to key policy documents of the New Labour era, however, is that the concept of mix is presented as a sort of cure for all sorts of urban ills, enabling social cohesion and sustainable communities. Against this background, achieving a stronger social mix was seen as key to reducing 'neighbourhood effects' and struggling social exclusion in areas of concentrated deprivation through the attraction of higher-income households. In practice, the main concern of this approach, however, was not so much to protect low-income households from displacement due to gentrification or to make sure that areas dominated by upper-income households got their fair share of poorer households, but rather to make sink estates and other poor areas more attractive for the middle-class.

How were these concepts implemented? Which major changes did New Labour undertake in the field of housing and urban renewal politics, and which new instruments were used? The guidelines for New Labour's housing policy were laid out in a Green Paper titled 'Quality and Choice' (DETR 2000), and the emphasis on 'choice' (if read against more traditional conceptions of housing as a 'right' within Old Labour) in the headline already provides insights into some of the new directions of the New Labour policies. In a nutshell, the most important changes advocated for in the paper were a restructuring of social and council rents, an increase in large-scale stock transfers and new policies to support ownership (Starter Home Initiative). At the same time, both Right to Buy and the deregulation of rent controls were to be continued, as they were seen as successful in supporting ownership. Thus, three main pillars of the Thatcherite revolution in housing politics – the sale of council housing, support for homeownership as the dominant tenure and the weakening of tenants' rights – were not reversed but maintained and even advanced. The new direction envisaged in the Green Paper 'buil(t) on the Conservatives' legacy and [took] it a stage further' (King 2001, p. 155). This is especially obvious with the large-scale stock transfers. While around 400,000 homes were transferred from councils to RSLs in the decade following 1988, New Labour aimed to decisively speed up this process and set a goal of transferring of up to 200,000 dwellings each year. Believing that transfers would be efficient in raising private capital (thus enabling the renovations needed to catch up with the £19 billion repair backlog that had built up in the council housing sector) and 'offer[ing] tenants a more diverse range of landlords', the clear goal was to sell off as many council houses as possible and to have RSLs provide the majority of social housing Along the way, this also redefined the role of local authorities, which New Labour saw less as providers of social services and more as having a 'strategic function' in choosing the most appropriate form of delivery for local needs and engaging in partnerships with the private and non-profit sector to push forward local development.

In a closely related move, arms-length management organisations (ALMOs) and private finance initiatives (PFIs) were introduced as alternatives for the management of council houses, for the cases where stock transfer initiatives had been defeated at the ballots.[5] While ALMOs are private companies set up by councils to manage the housing stock still owned by the council, PFIs are public-private partnerships with corporate banks, developers and consultants for regenerating and taking over the management of specific council estates. PFIs have become very controversial, and are blamed for escalating costs, delivering poor quality and undermining tenants' rights (see Hodkinson 2011a, 2011b). Stuart Hodkinson has even described PFI as a 'neoliberal straitjacket', intended to lock in gentrification-based regeneration at the neighbourhood level, guarantee long-term profits to (finance) capital and create powerful privatising and marketising pressures across the local public sphere (Hodkinson 2011a, p. 358). Both initiatives can be interpreted as a form of privatisation in which the state remains the owner of the housing stock, but the everyday management is changed in a way that is more adapted to commercial calculations, resulting in cost cutting, bad service and a lack of accountability.

Choice also played a key role in the new policy of the Blair government with regard to social rent restructuring. Thereby, the new approach contained three essentials elements that were to be implemented over the course of 10 years: (i) social rents were to be fixed at a level of 30–40% below market prices; (ii) as rents in the RSL sector were on average 20% higher than in the council sector, rents there should be successively increased to reduce the gap; and (iii) in order to achieve more fairness, social rents should reflect the value of the property more closely, allowing for higher rents in well-located and/or high-quality housing stock. All three reforms resulted in rent increases and de decommodified social rents, i.e. brought them closer in line with the ups and downs of the market.

Regarding support for homeownership, the New Labour government maintained the principal position of its Conservative precursors, while putting more emphasis on enabling access to a mortgage and reducing the risks for low-income home buyers. In this respect, the Green Paper announced as a Key Vision for the 21st Century:

> Most people want to own their own homes at some stage in their lives. Our policies support sustainable homeownership – that is, where the owner can meet the long term cost of buying and maintaining a home.
>
> (DETR 2000, p. 16)

While the general goal of expanding homeownership was shared with the Conservatives, the actual problems addressed and the instruments designed were a bit different and focused more on helping less well-off households, or those in areas where house prices were high, to achieve homeownership through shared ownership and low-cost homeownership initiatives, as well as providing benefit assistance with the payment of the mortgage interest for owners facing difficulties with their mortgage payments. The existing system of mortgage interest tax reliefs was widely abandoned and replaced by direct subsidies. Summing up, this set of reforms supported the low end of the owner-occupation market and provided extra support for people on the threshold of homeownership. The downside of this has been an accelerated financialisation (Aalbers 2016) of homeownership, bringing buyers into intensified relations with, and dependency on, global financial markets.

Over the years, a number of instruments have been put in place to implement this ambition. Thus, partial ownership programmes have been greatly expanded. These enable a household to purchase a proportion of a new dwelling provided by a housing association and rent the rest, with the right to upgrade to full ownership. Shared ownership was first introduced by the Conservatives in the 1980s and then expanded greatly with the help of the HomeBuy programme in which households could buy 75% of a property with a market rate mortgage and 25% with a zero-interest equity mortgage. In a similar way, the Starter Home Initiative (later Key Worker Living Programme) enabled key workers in the notoriously high-priced southeast of England to buy as little as 25% of a property from a housing association and rent the rest, thus, enabling health workers, teachers and police officers to buy a house (and take a job) in London. In addition, ownership was also supported through a massive shift in the tenure composition of new developments. As (the above discussed) policy changes had severely limited the capacity of local authorities to subsidise social housing, new developments were dominated by market rate flats in owner-occupation to a higher degree, so the overall share of owner-occupation increased.

Summing up a plethora of policy initiatives (Imrie and Raco 2003, pp. 14–16, list altogether 151 policy programmes with relevance to urban policy), not only did New Labour adopt central features of the Conservative policies of the 1980s and 1990s, but it even accelerated the commodification of housing in a number of respects: it intensified the privatisation of council housing stocks and its transfer to housing associations, brought social rents more in line with the market, spread owner-occupation further down the income scale and opened up public housing stock to concerted gentrification.

Austerity and New 'Class War Conservatism' Under the Coalition Government

Since the 2010 elections, which brought a coalition of Conservatives and Liberal Democrats into power, the main direction of the housing policy inherited from New Labour has seen both a continuation and an intensification.

The ideological base of much of the new government's policy in the first years has often been summarised as 'the big society' and understood as an initiative towards strengthening social enterprises, community self-help, voluntarism and philanthropy, thus, providing an alternative to the 'big state' of Old Labour and the extreme individualism of the Thatcher years. Drawing on an interesting blend of conservative communitarianism (often referred to as 'Red Toryism' in the British press), libertarian paternalism and neoliberal hostility towards the state (for more details see Corbett and Walker 2012), big society concepts have aimed at strengthening active citizenship. In practice, however, the attempts to introduce new democratic rights to communities have been modest at best. In fact, reacting to the financial crisis of 2008, big society was successively replaced by austerity, and the Cameron government introduced the toughest budget cuts experienced in Britain since the 1920s, even larger than those seen under the Thatcher government. Thus, while the rhetoric and the philosophy of Cameron's government could be interpreted as an attempt to 'humanize Thatcherism' (Corbett and Walker 2012, p. 487) or be read as 'Tories' accommodation with Blairism' (Lowndes and Pratchett 2012, p. 36), in terms of practical policies, the main impact of his government in the field of housing was a vigorous policy of austerity and a more intensified continuation of market-oriented policies.

Thus, the Housing Strategy for England (HM Government 2011) that was published in November 2011, mainly portrayed the housing crisis in the southeast of England as a crisis facing aspiring homeowners and as an underachievement in building new construction. It put the blame on 'central planning, top-down targets and bureaucratic structures'. Consequently, the solution suggested was to lift the red tape and deregulate planning. The same line of argument was taken in the Localism Act (2011) and the Planning and Housing Act (2016). The following initiatives were new to the agenda (see Bowie 2017):

- The introduction of a streamlined National Planning Policy Framework (NPPF) with reduced targets for brownfield redevelopment and affordable housebuilding. In this context, the NPPF introduced a bias in favour of development, which required local authorities to prove that a development project did not comply with adopted planning goals, if they were to put regulations on it. Moreover, existing regulations on permitted developments were extended to allow developers to convert offices and industrial buildings into homes without a requirement for planning consent.

- A new focus on viability was been introduced; if developers could demonstrate that it is not profitable for them to build a new development that met the council's planning policy with regard to the delivery of affordable housing, they could request that these requirements be reduced or waived.

- In order to support households at the threshold of ownership, a new First Buy subsidy programme provided a 20% equity loan for first-time buyers, and a New Build Indemnity Scheme (NBIS) even offered 95% mortgages backed by a state guaranteed indemnity fund for mortgage lenders. Investors were also given more freedom to challenge local infrastructure and housing requirements and to apply for subsidies from a new Get Britain Building Investment Fund for stalled sites.

- Private and social housing tenants, in contrast, were punished rather than supported. Benefit levels for private rentals were reduced, and an absolute benefit cap at a rent level of £350 per week for single adults and £500 per week for couples was introduced. Especially in London, where rents are very high, this has resulted in an extreme acceleration of pressure on low-income households to move to more affordable locations. There is wide consensus that this has resulted in the poor being pushed out of Central London (see Hamnett 2010), or as the then mayor of London, Boris Johnson, put it more bluntly, a 'Kosovo-style social cleansing' of the capital (see http://www.bbc.co.uk/news/uk-politics-11643440, 28 October 2010).

- Funding for new social housing was also cut by over 50% and redirected to a new affordable rent model that increased the caps of the affordability definition to 80% of market rents and introduced flexible two-year tenancies (in contrast to the existing lifetime tenancies) to social housing.

- Most controversial was the introduction of a 'bedroom tax' for social tenants. In short, the new regulation meant that one's housing benefit was reduced if the home of the claimant was under-occupied. Thus, the 'eligible rent' is reduced by 14% for one extra bedroom and 25% for two (or more) extra bedrooms. Mostly referred to as the 'bedroom tax' by the public, in more official terms it is called 'size limit rules' or 'under-occupancy rules'. Whereas it might seem fair not to support under-occupation, the main problem with the bedroom tax is that under-occupying households in the social housing stock can, in practice, hardly find an affordable alternative on the private market. The bedroom tax, thus, increases the pressure on long-time council residents without providing them with many alternatives.

- Right to Buy was revitalised and discounts were increased from £16,000 in some areas to a maximum of £75,000 across England and £100,000 in London (to reflect higher property prices).

In light of such policies, it is easy to understand why Hodkinson and Robbins (2013, p. 59) described the Coalition government's housing policy as 'class war conservatism', aiming to 'unblock and expand the market, complete the residualisation of social housing and draw people into an ever more economically precarious housing experience in order to boost capitalist interests'.

Along with new planning and housing legislation, massive cuts in central government funding to local authorities had powerful effects. Reacting to the financial crisis of 2008, austerity became the official policy of the national government in 2009, resulting in ferocious public sector cuts, as well as cuts to social services, benefits and tax credits. Downsizing public expenses massively affected municipal budgets and led many cities to turn to desperate measures to fund frontline and statutory services or structural investments. The consequence was an increased pressure on generating new revenue streams from the sale of public land and social housing estates for redevelopment. All across the UK, but particularly in London, this has led to initiatives aiming to demolish existing social housing estates and reconstructing the higher densities in partnership with private developers. Modernist estates like the Heygate or the Aylesbury in Southwark in London have become subject to state-led gentrification and aggressive displacement (see Lees 2008, 2014; Lees and Ferreri 2016; Watt 2009a) and social housing is under intensified pressure to be privatised.

Conclusion: Neoliberalism, Tenurial Transformation and Gentrification

The UK has been, over the last 100 years discussed here, a paradigmatic example of what Kemeny (1995) has called a 'dualist market'. Homeownership is the linchpin around which the whole system of housing production and consumption has been developed in this country, and over time this basic orientation has become even stronger. In the UK, private rental housing is seen as a residual form of tenure, as a sort of 'waiting room' towards ownership and, therefore, only weakly regulated. While past governments had at least introduced some regulation, these regulations were abolished in the 1980s, and the sector has been widely deregulated. As a consequence, private tenants have hardly any sort of protection against the perils of the market. Protecting the lowest strata of the population who cannot afford to buy into ownership is made subject to the provision of social or council housing, i.e. to a sector that is separated from the rest of the market, widely decommodified and run by state or non-profit organisations.

Table 3.1 shows how the share of different tenures among households has developed in the London over the 55 years 1961–2016.

The history of British housing policies is clearly reflected in this table. Starting with private rented housing, this form of tenure was in a massive

TABLE 3.1 Trends in household tenures, London 1961–2016 (Based on Housing in London, 2021)

	1961	1971	1981	1991	2001	2011	2016
Owner-occupied	36.3%	40.4%	48.6%	57.2%	56.5%	49.5%	49.1%
Social rented	18.2%	24.9%	34.8%	28.9%	26.2%	24.1%	22.9%
Private rented	45.5%	34.8%	16.6%	13.9%	17.3%	26.4%	28.0%

decline until the 2000s but saw quite a spectacular revival in the next decade. New data suggests that this trend is ongoing. Social rented housing was greatly expanded by Labour governments until 1979. Since the Thatcher government, and continuing with both the New Labour and the Coalition governments, social housing has been under attack and the share of households living in this form of tenure decreased by around one-third. Added to this, social rents have increased and security of tenure has been undermined by a series of reforms implemented since the 1980s. The owner-occupied sector has been supported through this period and finally gained dominance in the 1990s. Since 2001, its share has been getting smaller, due less to a loss of units and more to the growing share of the private rental sector. Data from the 'Housing Trailers to the Labour Force Survey' (https://data.london.gov.uk/dataset/housing-london/resource/951398a1-fe73-462e-bb1d-a14155ae7d7f) shows that, within the owner-occupied segment, the share of outright owners has continuously grown since 1981, whereas the share of owners with a mortgage has declined. This reflects both the massive growth of this sector in the past, as well as the difficulties potential home buyers face today in an environment of inflated housing prices. In short, whereas in the past people were pushed from the private and social rented sector into homeownership, they are now pushed from homeownership into the private rental sector.

What are the consequences of this constellation for the functioning of gentrification? Starting with private renting, the situation in this sector has changed fundamentally over time. Until the 1980s, the private rental sector was subject to more or less extensive rent regulations that, together with less favourable taxation rules, made letting a comparatively less attractive business (I will return to this in more detail in Chapter 4 where I discuss gentrification in Barnsbury). Price increases in this sector were limited to some degree and the target groups for private rentals were poor, discriminated against or transitory households, rather than the middle or upper classes. After the deregulation introduced in 1988, the private rental sector has become almost an ideal type of free market. Rents can be negotiated freely, there is no security of tenure and evictions are possible within a period of two months. In short, where landlords see a chance to demand higher rents, they face no barriers to realising them. This laissez-faire environment has made renting an attractive investment and has supported a large growth of buy-to-let schemes that have expanded greatly over the last two decades. While the conditions of living in a private rental are rather unfavourable to many tenants, and have become worse over time, the sector has massively expanded. More and more households that would have traditionally aimed for homeownership find themselves stuck in private rentals. Instead of being a stepping-stone towards homeownership, private renting has become a very insecure normality. The growth of this sector, together with its deregulation, works in favour of gentrification; rents are only subject to the interplay of supply and demand and if there is a chance to achieve higher revenues through fixing up a property towards its highest and best use, getting rid of the existing residents and replacing them with tenants who can pay more, there is no barrier to achieving this goal. Economic displacement is the norm here. As a consequence, within the British context, a growth of the private rental sector means making gentrification very easy to achieve.

The situation is a bit more complicated with homeownership. First, home buyers are much more closely tied to land and financial markets than tenants. Buying a house is normally the biggest investment of a household in one's lifetime and it can usually only be done by using a mortgage. This ties home buyers to the financial sector and makes it very vulnerable to changes in the financial markets. This is especially the case in the UK, where variable interest rates and low limits on loan-to-value rates are the norm. At the same time, the type of property, location and other factors play an important role in relation to the household income and in making the investment viable. Thus, whereas ownership provides a form of security unmatched by any other form of tenure in the UK, it cannot be realised by all, and achieving this dream makes households vulnerable to changing financial conditions. In other words, there is a close interrelation between gentrification, rising home prices and the exclusion of social groups from achieving this form of tenure. This is exactly the case that can be seen in large parts of London, where a large swath of potential first-time buyers finds it impossible to acquire a flat they can afford. At the same time, in owner-occupation, the role of the investor (owner) and the consumer (occupier) are conflated (Hamnett and Randolph 1988, p. 73). A home owned is not just a place to live in, but also an asset. Homeowners (especially outright owners) can, thus, profit from rising prices, as these rises increase the value of their home. Due to this situation, owner-occupiers can put the focus on their different roles at different times in their lives and can enjoy increasing home prices when they have paid back their mortgage, as it adds value to their investment. This becomes particularly important when owners wish to convert the stored-up value of their property into cash (see Lowe 2011, pp. 189–194). The policy changes I have described, therefore, had a rather contradictory affect on owner-occupiers. On the one hand, they enabled a growing part of the British society to profit from price increases and made the gentrification of inner city neighbourhoods a business enjoyed by many; on the other hand, this business can only work as long as house prices keep rising, thus, excluding more people from owner-occupation and pushing them into the insecure private rental sector.

Social and council-owned housing has been 'the only de facto buffer against gentrification' (Watt and Minton 2016, p. 210; see also Watt 2009a, 2013) in the UK for a long time. Both the land and the houses built on it were owned by the public, rents were determined by administrative procedures and access to this stock was dependent on social needs and the length of time spent on the waiting list. Until sold, this form of housing was not subject to market processes and, consequently, the concept of a potential ground rent to be realised by the highest and best use, i.e. the rent-gap hypothesis, would not be applicable to it. The subjection of this land to market rules is only an outcome of deliberate political decisions. In this respect, both the Right to Buy, Housing Stock Transfers, and current regeneration schemes based on demolishing council housing and selling the land to an investor to redevelop it have produced 'state-induced rent gaps' (Watt 2009b, p. 235), i.e. opportunities for exploiting the enormous capital accumulation potential piled up in the local authority's housing stock. The ongoing decline of the social rented sector, thus, goes hand in hand with the progression of gentrification and 'the RTB [right to buy] in London swiftly morphed from a property-owning democracy into . . . a private landlord owning plutocracy as increasing numbers of ex-RTB properties were bought up by absentee landlords' (Watt and Minton 2016, p. 208). The comparably high share and its wide geographic distribution of public housing built until the 1970s, thus, have opened up immense investment opportunities through resales and buy-to-let schemes, which by now are even valorised by global financial investors.

The downsizing and 'residualisation' of social housing has moreover contributed to increased gentrification pressure elsewhere, as the lack of affordable housing has pushed more people into alternative forms of tenure, thus adding both to the inflation of housing prices in the owner-occupation sector and assisting the growth of the private rental sector.

Summarising the *longue durée* of British housing policies, it is evident that the changes implemented by subsequent governments have resulted in linking housing ever closer to the market, stimulating investment and making displacement easier. In effect, the policy changes described in this section have, therefore, all fostered gentrification and made it a blueprint of urban development strategies in general.

The German Experience

Compared with Britain, the two most obvious characteristics of the German housing system are its form as a rental system (instead of one designed around owner-occupation) and the long continuities that have determined its recent shaping. With the exception of the socialist interregnum on the territory of the German Democratic Republic (GDR)[6] (1949–1990), the fundamental constellations of Germany's housing policy were established in the 1920s and have since then have changed only gradually (see Schulz 1986; Novy 1990; Jaedicke and Wollmann 1991; Führer 1995; Mayer 1998; Bartholomäi 2004; Egner et al. 2004; Egner 2014). Core elements were developed under a system of a controlled housing economy (*Wohnungszwangswirtschaft*[7]), which was effectively in place from 1914 to 1960. In this context, existing rental contracts could be cancelled, rental prices were set, and the occupation of flats was controlled by the state for the most part. Only through successive decontrol, introduced step by step since the early 1960s, was market regulation able to gain more ground. Both the controlled housing economy and decontrol were accompanied by a system of state-subsidised social and non-profit housing. Later on, in the 1980s and 1990s, the marketisation of the housing sector was a central political project of the Conservative Christian Democratic Union (CDU), Christian Social Union (CSU), and Liberal Democrats (FDP) Coalition government, but was hardly pushed forward with the verve and energy known in its counterpart on the other side of the English Channel. Today, some of the pro-business changes introduced in the 1980s and 1990s are being brought into question again and reversed to some degree. In summary, while an orientation towards free markets has gained ground, it has also always been strongly contested, and neither market orientation nor social protection tendencies were able to gain complete dominance. In Germany, the pendulum between the two keeps moving back and forth, with successive governments usually correcting changes introduced by their predecessors, but without completely annulling them (see Seeger 1995; Egner et al. 2004; Egner 2014).

From the Controlled Housing Economy to the Lücke Plan

Before World War II, housing regulations in Germany were very similar to those in the UK, i.e. hardly existent. As in the UK, most housing was built by private developers and rented, and the housing market was more or less regulated by laissez-faire policies. Throughout the Wilhelminian period, Germany experienced a rapid urbanisation (Reulecke 1985), accompanied by a chronic imbalance between housing supply and demand, rising rents and deteriorating housing conditions for the working classes. As in the UK, this led to widespread discontent and an upswing of reformist and philanthropic movements. However, not much changed prior to World War I and one major reason for this was the aristocratic government structure of the German Reich (which was especially pronounced in Prussia, then Germany's largest state), which gave landlords more votes in local elections than property-less tenants. This effectively blocked all attempts towards reform and made tenants 'slaves' to the landlord (Führer 1995, p. 35).

Germany's entry in the war in 1914 came to crush this system. With the goal of strengthening the morale of German soldiers, the emperor introduced a moratorium on evictions on 4 August 1914. In practice, this meant that the families of German soldiers received assured tenancy and were impossible to evict. When housing production was diminished through war and rent prices went up, this security of tenure was more or less expanded to all households in 1917. Moreover, *Mieteinigungsämter* (departments for the settlement of rent disputes) were introduced and given the right to abolish the cancellation of existing rental contracts, determine rents, allocate tenants, stop the conversion of rental units to commercial uses and order demolition of unsound structures. These measures were drastic, but they were thought of as temporary and were justified as an emergency solution necessary to stabilise the homefront. However, the situation barely improved after the war had finished, and the deficit in the housing market grew due to marriages that had been postponed because of the war, returning soldiers and refugees from German territories lost after the Treaty of Versailles. In addition, Germany saw a full-blown revolution in 1918 that was followed by a political crisis, militant uprisings and a growth of communist movements. In this situation, a return to Wilhelminian style liberalism was virtually impossible. Thus, when the new Weimar Republic set up its first laws on rent regulations (most importantly the *Reichsmietengesetz* [law on rents], the *Mieterschutzgesetz* [law on security of tenure], and the *Wohnungsmangelgesetz* [law on the management of the housing shortage]), existing regulations were not only kept in place, but strengthened. The most important regulations introduced in 1922 and 1923 were the following (see Schulz 1986, pp. 140f.; Führer 1995, pp. 52f.):

- A cancellation of an existing rental contract through the landlord was made subject to a legal trial at the district court and could only to be justified in the case of: (i) disorderly conduct, (ii) rent arrears of two months or (iii) claiming the need of the apartment for the personal use of the landlord.[8]

- The legitimate rent was frozen at the level of the 'peace rent' that was in place on 1 July 1914 for all flats that had already been built at that time, and was set slightly higher rents for flats built during the war. Newly built homes were exempt from this regulation. In practice, this meant that rents were fixed at a very low level for most households, especially when inflation is taken into account.

• The management of housing through public departments was supposed to be abolished by 1933. However, as I discuss later in this section, the Nazis returned to it in 1936.

It hardly comes as a surprise that these regulations were strongly criticised by landlords and the right-wing establishment, both of which attacked the reforms as a form of 'housing bolshevism'. Interestingly, the industry was more ambivalent, as rents had often been used as a justification for demanding higher wages by the trade unions, meaning low rents could also be read as enabling low wages.

An important point about these regulations was that they were limited to buildings constructed up to World War I. New buildings were not part of this legislation and, therefore, a dual rental market developed in the 1920s in which the prices for new buildings were considerably higher.[9] Yet, new building was difficult in general. Inflation was severe and this resulted in a lack of accessible mortgages, so investors had tremendous difficulties accessing capital. This situation led to the emergence of new suppliers and the development of a system of public subsidies for supporting the construction of 'residential buildings' (*Objektförderung*)

To start with the latter, the German government(s) started covering, or at least subsidising, the calculated deficit on the normal profitability of private construction measure (*unrentierliche Kosten, Aufwendungszuschuss*) in return for more affordable rents and occupational rights. This was the birth of social housing in Germany in the form in which it has survived until today. As public subsidies accounted for 71% (1921) to 99% (1923) of all construction schemes (Schulz 1986, p. 147), the relevance of this policy can hardly be overestimated. As in the UK, the German state not only started to regulate housing prices, but – in the face of market failure – intervened into its actual production. There are, however, two fundamental differences between the kind of social housing introduced in Germany and council housing as it was invented in the UK. First, subsidies did not discriminate between public, non-profit and commercial providers in Germany, but were given to all applicants. As a consequence, a major part of social housing was actually built by commercial providers. Second, and closely related to this, the money spent did not result in a decommodified housing stock managed by public authorities; the regulations were seen as the price to be paid by an investor for public support and, thus, expired after a while.

Even with this support, raising capital remained a difficult issue and new construction remained far below pre-war levels. Consequently, a number of new institutions were founded to collect money, reduce buildings costs or provide cheap credit. The most important innovation in this respect was the foundation of charitable housing companies (*Gemeinnützige Wohnungsbaugesellschaften*). Called 'homesteads' (*Heimstätten*) at that time, they were:

> . . . commercial enterprises, but with state shares. They should assist the technical advancement of construction practices; organise the acquisition of building materials, land, and capital; and participate in local charitable housing construction . . . The major goal was to construct good, but inexpensive, small flats and thereby overcome a deficit left by the private housing market.
>
> (Schulz 1986, p. 146, own translation)

These housing providers were complemented by state-owned mortgage banks (most notably the *Deutsche Pfandbriefbank* owned by the Prussian government), building societies (*Bausparkassen*), faith-based or union associated housing providers and numerous cooperatives. All these institutions have survived in one form or another until today and, thus, as Bartholomäi (2004, p. 20, own translation) put it, 'the relevant institutional landscape [of German housing policies] originates in the 1920s'.

In summary, the pillars of the German housing system, as they were erected in the 1920s, were meant as a makeshift solution – yet, one clearly leaning towards the interest of tenants. This was decisive for the future:

> Regulating the misery was originally an authoritarian emergency solution. Yet, with the end of the monarchy, political parties came to power that had demanded state intervention in the housing sector previously. Now it became, and stayed, part of the welfarist credo that the state would be responsible for a sufficient provision of housing, as well as for affordable rents. Vested rights [*Besitzstände*] were established in housing policies that were hardly revisable.
>
> (Schulz 1986, p. 163, own translation)

When the Nazis came to power, they by and large kept with the existing housing constellation, but modified it to serve their racist policies and support war preparation. In this context, rent control was tightened and extended to all households, with the exception of Jewish citizens. Public housing control (which had been abolished only for a short time by the Brüning government) was also reintroduced and strengthened, first with regard to the use of residential space for commercial purposes (1936), later for the provision of housing to child-rich families (1939) and expanded to a system of complete public housing allocation when Allied air raids resulted in the severe loss of housing during the war. Subsidies for private construction were massively expanded and construction activities in fact increased until 1937. After this, however, means were directed towards war preparation and later to war activities, and construction figures plummeted again.

When World War II ended, the deficit of housing was bigger than ever before. The halting of building activities during the war, the serious losses due to aerial warfare by the Allies, and the millions of displaced persons resulted in a blatant housing shortage. In 1945, around one-fifth of the existing housing stock was destroyed (Kofner 2004, p. 152) and the housing deficit was estimated at 4.8 million units – relative to a stock of 9.5 million (Schulz 1986, p. 157). The situation was even more difficult in many big cities, such as Kiel, Hamburg, Bochum, Essen, Köln, Düsseldorf, Mainz and Dresden, where more than 60% of the housing stock was destroyed. To make matters worse, the banking sector was very unstable throughout the second half of the 1940s, making mortgage provision extremely inaccessible. In sum, the housing needs were severe and there were few prospects for improvement in the short-term.

In this situation, the Allied Control Authority had no choice but to continue and even intensify the existing controls on the construction, provision and pricing of housing. Starting in March 1946, all housing (in West Germany) was only to be rented through housing offices. These offices also had the right to billet families in need to excess private housing, and this was practised intensively. Rents were frozen at the 1936 level (which, in fact, means to the 1922 level for most apartments) and cancelling existing contracts was made next to impossible for landlords. In practice, 'regulations went so far and were so comprehensive that a housing market did not even exist in a rudimentary form anymore' (Kofner 2004, p. 153, own translation). The foundation of the Federal Republic of Germany in 1949 and the end of Allied authority didn't change much in this respect, and it took years until the first steps towards dissolving the existing controls were feasible and the system moved closer towards a market economy. In this respect, the major initiatives were the following:

- The First Housing Act (*Erstes Wohnungsbaugesetz, 1. WoBauG*) of 1950 liberated all privately financed new construction from occupation control through the housing offices.
- The First Housing Act also released privately financed residential buildings built after 31 December 1949 from price control and enabled market prices in this sector. Additional opportunities for rent increases were granted with the First Federal Rent Law *(Erstes Bundesmietengesetz, BMG)*.

- The Law on Housing Management (*Wohnraumbewirtschaftungsgesetz, WBG*) expanded the right of landlords to choose among the applicants entitled to a flat.

Given the low level of construction activities, the de-freezing of existing regulations, however, only applied to a very minor share of the stock. Moreover, these first steps towards deregulation were accompanied by supporting and strengthening the social housing sector. Together with the First Housing Act, a new legal framework for social housing was designed, and this was done in a way that was closely aligned with pre-war policies. The role of the state was again defined as providing subsidies for reducing the capital expenditures of private investors, in return for which rent caps (*Richtsatzmiete*) and occupational obligations could be demanded. These basic principles were strengthened with the Second Housing Act (*Zweites Wohnungsbaugesetz, 2. WoBauG*) of 1956, which remained in place until 2000 and came to be the major legislation for social housing throughout the history of modern Germany. Instead of fixed rent caps, it introduced cost rents (*Kostenmieten*) that reflect the costs of the investor as a basis for subsidies.[10] At the same time, the Second Housing Act defined income limits for the eligibility for social housing. However, these were fairly lax, and the goal of social housing was defined as providing adequate housing to a 'broad strata of the population' (*breite Bevölkerungsschichten*).

The second novelty to come with the Second Housing Act was a stronger orientation towards supporting owner-occupation, which was put legally on an equal footing with the support of rental housing. Very much based on the conservative ideology of the ruling Christian Democrats, supporting homeownership was declared a necessity for 'banding together wide parts of the population with property, especially in the form of family homes' (Second Housing Act, *2. WoBauG*, §1, own translation). While this justified extensive tax deductions, direct subsidies for owner-occupation were always minor as compared to subsidies for social housing.

Altogether, housing production was strongly supported by the West German state, which provided enormous subsidies in the 1950s. This resulted in the construction of an impressive number of new homes and, as a consequence, a large part of the inherited housing problem was overcome quickly. In the year 1960, the housing deficit had already decreased from 62.9% to 16.6% of the total stock (Mayer 1998, p. 432). This remarkable success was mostly an outcome of the very 'visible hands' of the state, which invested heavily in social housing. As a consequence, new housing was to a large part social housing: close to two-thirds of the 7 million new units constructed between 1950 and 1962 were social housing units (Kofner 2004, p. 157).

The more the immediate housing shortage was overcome, the more the ruling Conservatives saw the time as ripe for ending the controlled housing economy and introducing real markets. This was eventually done with the Law on Decontrol and Social Rent and Housing Legislation (*Gesetz zum Abbau der Wohnungszwangswirtschaft und über ein soziales Miet- und Wohnrecht*) of June 1960. Named after the minister for housing construction, the law is usually referred to as the 'Lücke' Plan in the German literature. In short, the Lücke Plan abolished the existing system of housing allocation through public housing offices and deregulated rent prices in all cities and counties where the housing deficit was less than 3% of the total stock. These cities were referred to as 'white circle(d)' (*Weißer Kreis*). The deregulation of rents started in 1963, but it was delayed twice and it took until 1974 in the case of Hamburg, until 1975 in the case of Munich and even until 1988 in the case of West Berlin (Welch Guerra 2016) until it was finally implemented. Until that time, rents in flats built before 1948 (and this was the lion's share in most cities) were regulated by local guidelines.

In order to cushion the introduction of market principles, housing allowances were introduced and protection against eviction was strengthened. Moreover, (regulated) market rents would not apply to social housing, where rent prices were subject to specific guidelines.

With this, the basic architecture of the German housing system that has survived until today was set in place. It is worth taking a step back and looking at the ideological foundations of this system. The conceptual backbones of the German housing system can be found in the social market economy model (*Soziale Marktwirtschaft*) that was promoted and implemented by the Christian Democrats in the 1950s and is still crucial for understanding Germany's character as a conservative welfare state. Strongly inspired by ordoliberal ideas developed by German economists and legal theorists (most notably by Wilhelm Röpke, Walter Eucken and Alfred Müller Armack) and Catholic ethics, the concept of a social market economy can be described as a combination of a free market with social welfare. In effect, the ordoliberals argued for designing the economy in such a manner that social goals were built into the market. State intervention was seen as necessary for establishing fair competition and maintaining a balance between economic growth and welfare, but it should be designed in a market-conforming (*marktkonform*) way (see also Kemeny 1995; Mau 2003; Stedman Jones 2012; Goldschmidt 2013). Ordoliberals saw the social market economy as a third way that provided:

> a [neo-]liberal alternative to laissez faire liberalism and collective forms of political economy, ranging from Bismarckian paternalism to social-democratic ideas of social justice, from Keynesianism and Bolshevism.

> In the face of Weimar mass democracy, economic crisis and political tur-
> moil, they advanced a programme of liberal-conservative transformation
> that focused on the strong state as the locus of social and economic order.
> The dictum that the free economy depends on the strong state is key to
> its theoretical stance.
>
> (Bonefeld 2012, p. 633)

Two principles are crucial to the social market economy: solidarity and subsidiarity. The principle of solidarity entails a responsibility of the individual towards the society, as well as a responsibility of the society towards protecting and assisting its members. The German constitution, thus, comes with a clear protection of private property, but it also underlines that 'Property entails responsibilities. Its use shall serve the public good' (§14 (2), *Grundgesetz der Bundesrepublik Deutschland* [Constitution of the Federal Republic of Germany]). The consequence of this is an obligation towards protecting private property and economic interests, but also a justification for state intervention. The second principle is subsidiarity demands that state intervention should have a subsidiary character and only be allowed where private actions fail. It, thus, has to be limited in reach and time, justified by market failure, and it needs to be designed in a way that wouldn't stand in the way of reintroducing market mechanisms.

It is easy to recognise these principles in the reforms described above. While in general, the state had an obligation towards enabling a market economy and overcoming the established controls on housing, it was also obliged to intervene in the interest of the public good if the market failed to provide adequate housing. This intervention would, however, needed to be subsidiary, i.e. limited in time, place and subject, and be designed in a way that would not contradict the proper workings of markets. As we will see in Chapter 5, this swing between social responsibility and market liberalism gave birth to ongoing oscillations in German housing policies that have defined the effort to find a balance between social responsibilities and economic efficiency very differently at different times, but still within the foundations laid out under the chancellorship of Konrad Adenauer.

The Design of Tenant Protections

On this basis, the typical pattern of German housing policies established since the 1960s has been a 'moving consensus' between Social Democrats and Christian Democrats (see Jaedicke and Wollmann 1991). In this environment, the basic constellations stayed intact, yet considerable parts of the consensus remained contested and subsequently changed in accordance with the orientation of the government in power.

In this respect, a first set of changes was introduced when a coalition government between Social Democrats and Liberals came into power in the early 1970s. The more the Lücke Plan was progressing, the clearer the lack of protection for tenants in the private rental sector became. In essence, the Lücke Plan effected the setting of the balance of power between tenants and landlords back to the state of affairs before World War I. The expansion of white circles equated with rent increases and, translated into the world of politics, this meant that more and more voters became critical of the liberalisation policies that were pushed forward.

Against this background, the new Social Democratic government took action and introduced a First Law for the Protection of Residential Tenants Against Eviction (*Erstes Wohnraumkündigungsschutzgesetz*, 1. *WKSchG*) and a Law for the Improvement of Rent Legislation and the Restriction of Rent Increases (*Gesetz zur Verbesserung des Mietrechts und zur Begrenzung des Mietanstiegs, MietVerbG*) in 1971. Along with this 'comeback of the interventionists' (Mayer 1998, p. 191), came three core innovations which (with the changes detailed below) are still in place today:

- An asymmetric termination of contract protection was introduced that benefited tenants more than landlords. Thus, tenants could terminate existing contracts without determining the reasons for doing so within a period of two to four months. Landlords were only allowed to terminate an existing contract for a short list of clearly defined reasons. The most important of these were rent arrears and claiming the property for personal use (*Eigenbedarf*).

- Rent increases were only allowed in line with the typical rent of the area (*Vergleichsmiete*), usually defined by a rent index (*Mietspiegel*, literally 'rent mirror') that would provide the average rents paid for a specific quality of building, with specific facilities, in a specific area. Moreover, rent increases could only be demanded periodically and only up to a certain amount above the original rent.

- Demanding a rent more than 50% above the level defined by the rent index (e.g. in the case of a new contract) would be regarded as extortionate (*Wucher*) and subject to criminal persecution.

With this, the Social Democrats introduced a nationwide system of tenant protections. The major characteristic was that it would enable rent increases, but cap rent adaptation at a level in line with the general developments in the market (see also Kofner 2004).

This system of tenant protection was partly softened when the Second Law for the Protection of Residential Tenants Against Eviction (*Zweites*

Wohnraumkündigungsschutzgesetz, 2. WKSchG) was passed in 1974. Against the background of maintenance deficits in large parts of the existing housing stock and high interest rates, it introduced a modernisation fee (*Modernisierungsumlage*) of 14% (since 1978, 11%; since 2017, 9%) with regard to the costs of measures 'capable of increasing the use value of a flat' (*gebrauchswerterhöhende Maßnahmen*). By unilateral declaration, landlords were, thus, entitled to add 14% of all modernisation costs (exclusive of financing and maintenance costs) to the yearly rent, and tenants had an 'obligation to tolerate' (*Duldungspflicht*) this (or move out). The relevance of this legislation is best explained with a simple example: if a flat without central heating cost 300 DM and the landlord decided to install a modern heating system for 6000 DM, the rent would be increased by 14% of this amount annually, that is 70 DM per month. Usually this would be way above the range of typical rents for the area. Modernisation, thus, enables landlords to leapfrog the existing rent restrictions and quickly arrive at a higher rent level.

This became more important when tax deductions (which had been in place for new construction since 1949) were expanded to existing buildings in 1978. In contrast to direct subsidies, e.g. those granted for social housing, tax allowances came free to the investor, as they were not linked to price caps or occupational obligations. Moreover, they stimulated expensive cost measures, thus providing additional benefits to lift up residential homes to their highest and best use. Both the opportunity to let tenants pay for improvements done to the flats they inhabit and the granting of tax giveaways for exactly these measures, made investment in existing residential buildings lucrative, and successively opened these stocks for reinvestment and gentrification.

The Conservative *Wende*

When a Christian Democrat government came into power in 1980, these reforms were kept in place. In contrast to the UK, the roll back aimed at by the conservative *Wende* (turnaround) promised by Chancellor Kohl did not break with the existing system, but rather radicalised its liberal components in a tentative and stepwise manner.

Although the Conservatives were strongly in favour of risking more market economy, only minor steps were taken to reverse eviction protections and rent regulations introduced by the Social Democrats a decade previously. Thus, for example, the requirements for local rent indices were changed in a way that would only include contracts settled in the previous three years, effectively narrowing the corridor for typical rents and enabling faster rent increases. Furthermore, landlords and tenants were allowed to bypass the comparable rent system and agree on predetermined rent increases (*Staffelmieten*). In practice, however, this was rarely accepted. To summarise, as the author of a PhD thesis mourned in 1998, 'the basic structures of rent price legislation and protections against evictions were not touched. At this point, the political limits to market-oriented strategies became apparent' (Mayer 1998, p. 204, own translation).

More important were attacks on the social housing and non-profit sectors. The Kohl government turned off the tap of public investment into social housing, as had its Conservative counterparts in the UK. The subsidies for social housing were reduced from 2.29 billion DM (1983) to 450 million DM (1988) within five years. At the same time, new funding schemes were established that enabled higher rents and owner-occupation for newly built social housing (*2. Förderweg*). Given the construction of social housing as a temporary subsidy, this predetermined a subsequent shrinkage of this sector. Thus, the less (new) social housing was subsidised, the more the total number of social housing units decreased.[11] It should be emphasised, however, that the decrease of social housing units was 'not an infamy of post-Fordist housing policies, or a national tragedy falling off the roof (as some contributions want us to believe), but a systematic consequence of the design of social housing construction as a subsidised, but still basically capitalist, form of private housing construction' (Becker 1988, p. 104). Compared with the Tories, one could argue, the German Conservatives had it easier, as they just needed to stop allocating money to see the social housing sector disappear.

The withdrawal from social housing was supported by the abolition of tax privileges for non-profit companies (*Gemeinnützige Wohnungsbaugesellschaften*) in 1990 (see Kuhnert and Leps 2017). As I've described, non-profit housing companies had a long tradition in Germany, dating back to the 1920s. Owned by municipalities, trade unions, charity foundations and cooperatives, they had been the backbone of affordable housing provision over decades and were supported through tax privileges. In return, they were restricted with regard to their profits, they had to limit their business activities to the housing sector, and they had to grant additional rights to their tenants.[12] With the abolition of the tax privileges, these companies – which owned 3.4 million flats at the end of the 1980s (Kofner 2004, p. 31) – were pushed closer towards the market. While many former non-profits maintained their previous management and letting policies, others were quick to implement rent increases and sold parts of their stock. The consequence was an intensified decrease of de facto social housing (see Bernt 2017) and a marketisation of the sector.

Summing up, whereas the Social Democrats had supported a growth of the social housing sector and implemented regulations that protected private tenants until the mid-1970s, this was partly reversed under the Schmidt and Kohl governments, and money streams were directed towards more market-oriented forms of housing provision. Thus, while the total volume of subsidies for social housing construction was reduced in the course of the 1980s, considerably more money was put into supporting homeownership through tax deductions as well as housing allowances, thus shifting subsidies from brick and mortar to individual support. Table 3.2 documents this development.

The more the nation state withdrew from financing affordable housing, the more important municipalities became in developing fragmentary forms of municipal counterpolicies. Two developments are crucial in this respect. First, in accordance with the three-tier design of German statehood, municipalities not only have a number of constitutional responsibilities for the provision of 'adequate living standards' in their territories, but they also hold the right to numerous decision-making powers. Against this background, municipalities have often come to work as a sort of repair service, addressing the market-oriented reforms of the national government. This is especially the case in big cities, many of which have Red or Red-Green[13] political majorities. Second, in the context of urban renewal, Germany has introduced numerous instruments that provide municipalities extraordinary planning rights. Since the 1970s, more and more cities have started using the declaration of Urban Renewal and Milieux Protection Areas as a means to implement local regulations on the development of rent levels. This will be discussed in more detail in Chapter 5, and it will suffice to state here that the deregulation of housing policies at the national level was partly accompanied by a re-regulation at the local level.

TABLE 3.2 Subsidies for housing and housing construction by types of subsidies in billions of DM, 1965–1988 (Modified from Mayer 1998, p. 435, own changes)

Type of subsidy	1965	1970	1975	1980	1985	1988
Construction of social housing	4.5	2.1	4.7	6.8	6.8	5.5
Tax deduction for the support of ownership	1.2	1.5	3.3	5.5	7.2	7.1
Subsidies for renovating existing stocks	0.1	0.2	0.3	1.1	0.8	0.9
Housing allowances	0.1	0.6	1.7	1.8	2.5	3.7
Urban renewal	0.1	0.3	1.1	2.3	1.5	2.7

Reunification and Neoliberal Consensus

The collapse of state socialism and the subsequent reunification of Germany in 1990 resulted in the introduction of a number of very specific policies that were only in place in the former GDR. These aimed to manage the transformation from a planned to a market economy. As these are of importance for understanding the case of Prenzlauer Berg, the most important aspects will be discussed in Chapter 5. The most important changes can, however, be summarised as follows:

- The former state administrations that were responsible for all housing stock built under socialism (and the stocks taken over by the state between 1945 and 1990) were transformed into commercial enterprises owned by the municipalities. Following the Existing Debts Assistance Act (*Altschuldenhilfegesetz*) in 1993, they were obliged to sell one-sixth of their stock. The privatisation was originally intended to enable sitting tenants to buy their flats, but, in practice, most flats were sold to real estate speculators (see Borst 1996; Bernt 2017; Bernt et al. 2017).

- Houses built before 1949 and expropriated either under the Nazis or by the communist regime were to be returned to the original owners or their heirs. This restitution proceeded until the turn of the millennium, and usually resulted in the quick sale of the affected properties.

- With the goal of stimulating private investments, extensive opportunities for tax deductions were introduced in the territory of the former GDR, and until 1998, investors could write off 50% of their investment costs against taxes within four years.

- Until 1998, the existing very low rents were increased in five steps to a bring them closer to market levels. Together with this, the existing (West) German rent legislation was expanded to the territory of the former GDR.

In summary, while all these conditions were directed at adapting the housing sector in East Germany to the West German 'normality', they resulted in a very specific set of regulations and enabled specific dynamics that were unique to the housing market in East Germany. While these were supportive of gentrification in some cases, they also brought about a strong population decline and shrinkage elsewhere (see Glock and Häußermann 2004; Bernt 2009).

For Germany as a whole, policy changes introduced by the Red-Green government that took power in 1998 were more decisive. In this regard, the most important initiatives are explained here.

First, the government worked hard on ending the direct subsidies for achieving owner-occupation that had proven ineffective and very costly in the preceding decades. Seeking to consolidate strained budgets, the new government tried to eliminate this item, but was blocked in the *Bundesrat* (the federal assembly of German states), where the conservative opposition had a majority in the first place. Eventually, a new coalition of Social Democrats and Christian Democrats scrapped the subsidy in 2006.

Second, the Red-Green government ended subsidies for social housing in 2001. This was justified by voicing a peculiar mix of concerns about a perceived lack of social mix in the existing social housing stock (which is a line of argument well-known from other Western countries), references towards the dramatic population losses in East Germany and the resulting general oversupply of housing, and a new focus on austerity and the need for budget consolidation. As social housing is a time-limited institution in Germany that – without renewed subsidies – is doomed to expire after a while, ending this form of subsidy necessarily led to speeding up the depletion of the existing social housing stock.

Third, the Schröder government reformed rent legislation, expanding the basis on which 'typical rents' were to be calculated and reducing the rent increases for sitting tenants to 20% (instead of 30%) within three years.

Fourth, housing allowances, which had for a long time lagged behind the increase in housing costs, were adapted and brought in line with inflation. At the same time, together with the reform of welfare benefit regulations (*Hartz IV*), housing support for the unemployed was redefined as a part of welfare benefits and considerably reduced.

In summary, the changes pushed forward by the Schröder government were very much in line with the Third Way thinking popular among European Social Democrats at this time and aimed at less intervention and control, and more governance. As Egner (2014, p. 18, own translation) points out:

> These reforms set in motion a change from 'housing policy' [*Wohnungspolitik*] to 'housing market policy' [*Wohnungsmarktpolitik*]. Contrasting the 'good old times,' the state withdrew from its role as a provider of housing, respectively a supporter of providing housing, and limited its task to the support of tenants through allowances.

In other words, while keeping with the general setup of the German housing sector, the government leaned back and refrained from further intervention. Within less than 10 years, this led to an intensive housing crisis in most of Germany's big cities and university towns (Schönig 2013). In most big cities this stimulated a return of housing as a policy issue and most German states have started new social housing programmes or introduced local regulations on rent increases. This is most visible in Berlin, where a new Red-Red-Green government has introduced a variety of policies directed at regaining more control over the housing stock. These include the massive expansion of the municipal stock, but also new regulations regarding rent levels and other policies (more details are discussed in Chapter 5). At the federal level, housing policies are, however, stuck in limbo. While new rent legislation and new funding programmes for social housing have been under debate for years, not much has been achieved yet. The most paradigmatic example of this is the 2015 introduction of a new rent brake (*Mietpreisbremse*) that was to limit the rent for new contracts to a level 10% above the rental index. As the new law excluded intensively modernised flats, new construction and apartments, in which the rent in the previous contract had already been above the rental index, it hardly had much of an effect. This has been extensively criticised – yet without much effect. Similar stories can be told for other areas (e.g. the reform of rental indices, new subsidies for social housing and the taxation of non-profit housing providers). In all these areas, federal policy initiatives are hampered by a stalemate between the Social Democrats and Christian Democrats, who together form the current government. This situation has prompted municipal governments to develop their own housing policies and, at the time of writing, numerous new policy initiatives are under construction in cities like Berlin, Hamburg and Munich. It is still too early to see whether this will lead to a stronger fragmentation of housing policies between different localities, or whether the national state will re-enter the stage (and what policy approach it might adopt). Thus, while in general the neoliberal consensus of the Schröder era is hardly unbroken, there are signs of hope and it seems likely that the pendulum swing of German housing policies described will change its direction again.

Conclusion: Gentrification Between Regulation and Deregulation

Summarising a century of reforms, German housing policies can best be understood as long cycles of regulation, deregulation and re-regulation revolving around renting as the dominant form of tenure. Although a stronger orientation towards markets is unmistakable since at least the conservative *Wende* in 1982 (and many would even argue since the Lücke Plan in 1960), it has always operated as neoliberalisation with brakes on. Changes in the general setup of the housing system have remained limited, and modifications introduced by new governments were focused on fine-tuning rather than overturning existing regulations.

The policies towards rent regulations for sitting tenancies provide a vivid example of this. As the majority of German households rent their housing (in Berlin more than 85%), rent regulations are of major importance for German housing policies in general. Table 3.3 lists the most important changes in this field between 1971 and 2019. As it demonstrates, the overall design of the regulations has seen a lot of back and forth, but with little fundamental change since they were introduced in the early 1970s.

TABLE 3.3 Regulations on rent increases in sitting tenancies in Germany, 1971–2019 (own composition)

Name of the law	Government under which the law was introduced	Regulations
Wohnraumkündigungsschutzgesetz (WKSchG) 1971	Social Democrats and Liberal Democrats	Rents can only be increased up to a level of comparable flats typical for the area (*Vergleichsmiete*), based on a local rent index (*Mietspiegel*)
Zweites Wohnraumkündigungsschutzgesetz, 2. WKSchG, Miethöhegesetz (MHG) 1974		If flats are modernised, rents can be increased by a modernisation fee (*Modernisierungsumlage*) of 14% per annum
Wohnraummodernisierungsänderungsgesetz (WoModÄndG) 1978	Social Democrats and Liberal Democrats	Modernisation fee (*Modernisierungsumlage*) reduced to 11% per annum
Gesetz zur Erhöhung des Angebots von Mietwohnungen 1982	Christian Democratic Union, Christian Social Union and Liberal Democrats	Rent indices (*Mietspiegel*) to be defined on the basis of flats where rents have changed during the year preceding their drafting
		Rent increases (without modernisation) limited to 30% within three years
Mietrechtsreformgesetz 2001	Social Democrats and Green Party	Rent indices (*Mietspiegel*) to be defined on the basis of flats where rents have changed during the four years preceding their drafting
		Guidelines for the setup of qualified rent indices (*Qualifizierter Mietspiegel*)
		Rent increases (without modernisation) limited to 20% within three years

MietNovG 2015	Christian Democratic Union, Christian Social Union and Social Democrats	The rent for new tenancies is limited to a level of 10% above the rent of comparable flats typical for the area (*Vergleichsmiete*), yet new constructions and extensively modernised flats are exempt from this regulation
		A rent brake (*Mietenbremse*) is introduced that limits the rents in new contracts at a level 20% above the respective rental index. New buildings, intensively modernised flats, and flats where the previous rent was already above this level are exempt from this regulation
MietAnpG 2018	Christian Democratic Union, Christian Social Union and Social Democrats	Modernisation fee (*Modernisierungsumlage*) reduced to 8% per annum and limited to an amount of 3€ per m^2
		A duty to disclose the previous rent when starting a new contract is introduced, thus, closing one of the loopholes of MietNovG 2015

One can easily see how policy initiatives have revolved around three core parameters: (i) the extent to which existing rents are included in the definition of rent indices (*Mietspiegel*), (ii) the range of caps on rent increases within a timeframe, and (iii) the acceptable amount of modernisation fees as a share of total costs. On this basis, the rent to be paid by most German households is a hybrid of market rents and the average rent level, which makes rent increases possible, but also operates as a brake against them. How this contradiction is solved in practice is defined by a set of parameters that have been fixed for next to half a century now and have only gradually changed. As a rule of thumb, the actual benefits of these parameters goes to the landlords and investors when the Conservatives are in power and to the renters when the Social Democrats rule. The changes are, however, never excessive.

This structure is by no means contingent. It is rooted in the specificities of German statehood and, in this sense, the conditions for rent increases, respectively gentrification, reflect the *longue durée* of relations between state, capital, land and people in the 'rhinian' variant of capitalism.

The first issue to be discussed in this context is the above-sketched ordoliberal underpinning of public policies in Germany. Here, the influence of ordoliberal thinking was decisive for designing the concept of a social market economy, which is crucial for the design of public policies in Germany and has even found its way into the German constitution. In this model, the dominance of market forces in allocating goods is at the same time guaranteed and curtailed by the state. In line with this orientation, German policy-makers have always been reluctant to intervene directly in the housing market, on the one hand, e.g. by providing non-market housing but have developed a plethora of regulations and incentives for market participants, aimed at fine-tuning the working of the market, on the other hand. State intervention and market orientation are two sides of the same coin here, and markets are not allowed to operate freely, nor can state intervention work in a way that would severely contradict market principles in any area of housing. In summary, the model upon which the German housing system was developed is at the same time market-oriented and state-interventionist. It favours the market as a superior instrument for the efficient allocation of land, capital and goods, yet simultaneously rests on state-control and public responsibility.

The second specificity that sets the German political system apart from to the Westminster model of the majority rule dominating political life in the Anglo-American world, has been coined 'negotiating democracy', 'consociational democracy' or 'consensus state' by political scientists (Lehmbruch 1976, 1996; Lijphart 1977, 1999; Czada and Schmidt 1993; Czada 2000, 2003). In a nutshell, it reflects the unanimous assessment that political decision-making in Germany is to a strong degree directed at achieving agreement between major political players. Three major causes have been depicted for this (see for example Czada 2000, 2003). First, for a long time in history, political life in Germany has been fragmented between interest groups and milieux that gave rise to a comparably diversified political landscape. As a consequence, most governments in Germany are coalition governments of at least two parties, whereby constellations change over time. In this environment, political ambitions of any kind cannot grow without seeking compromise. This is especially the case when it comes to decisions that need to be processed within the three-tier system of government that is fundamental for German statehood and forms the second feature promoting consensus orientation. Here, the national government, the states/regions and municipalities all hold important veto powers. As the political majorities almost always differ in these three tiers, the interaction of different levels of statehood is made comparatively difficult. German political scientists have even talked of a 'joint-decision trap' (*Politikverflechtungsfalle*, Scharpf 1985). Third, resulting from both ordoliberal ideas and the historical strength of Social Democrats and trade unions, Germany has developed patterns of a 'sectorally segmented meso-corporatism' (Czada 1985) in which diverging interest groups are included in policy-making in a number of policy areas. Historically, housing policies have been a prime example of this and there is hardly a major policy initiative in this field that has not been accompanied by intensive communication between state bureaucracies, the housing business lobby and tenant organisations. Moreover, and partly resulting from this interlocking, the field of housing has strongly been influenced by expert communities that have formed across public and private organisations and different state levels. Together, all this leads to a dominance of policies of the middle ground, in which historical compromises strongly shape future trajectories of policy-making. In this environment, housing policies have been described, as mentioned earlier, as a 'moving consensus' (Jaedicke and Wolmann 1991, pp. 432–433) in which key decisions of previous governments are rarely revised, but instead are modified and altered by recent decision-makers. The political leeway for reform, no matter which direction, is limited, and once established, policies remain the point of reference for a long time.

What is the relevance of this for gentrification? As described, the outcome of the negotiating democracy is a complex set of national regulations that define the opportunities of rent increases according to different situations. These regulations, however, interact with each other and can be, moreover, added to local and sublocal regulations. The outcome is a complex system of positions and standard situations. In this sense, Table 3.4 sketches the most important regulations towards rent increases in Germany and Berlin.

As can be seen, the opportunities for rent increases in the private rental sector (which makes up the majority of the housing stock in all large German

TABLE 3.4 Regulations on rent increases in Berlin and Germany as of 2019

	Sitting tenancy	New tenancy	Modernisation	Conversion from rental to condominium status
National regulations	Rents can be increased up to the usual price for a comparable flat (*Vergleichsmiete*), usually defined by a rent index Rent increases are limited to an amount of 20% over three years	No regulation, in principle Rents at 50% above the usual price for a comparable flat (*Vergleichsmiete*), can be regarded as exorbitant and prohibited by a court	Rents can be increased by an amount of 8% of the costs for all installations that increase the use value of a flat (like modern heating, insulated windows, etc.) per annum, yet only to an amount of 3€ per m²	All regulations for sitting and new tenancies stay in place, yet security of tenure is strongly undermined, as owners can dismiss sitting tenancies by claiming personal use
Regulations in Berlin	(In operation from 2020 on, but revised by the Federal Administrative Court in April 2021) All rents are subject to a rent cap (*Mietendeckel*) and limited to an amount of 3.92€ (for unrenovated flats) to 9.80€ per m² New constructions are exempt from this regulation Modernisation fees are made possible up to a level of 1€ per m²	Rents should not be greater than 10% above the price for comparable flats (*Vergleichsmiete*) New buildings and extensively modernised flats are exempt from this regulation		Owners cannot claim personal use within a period of 10 years after the transformation of a rental flat to a condominium
Regulations in Milieux Protection Areas in Berlin			Luxury modernisations (e.g. marble ceilings, second balcony, etc.) are banned	Transforming a flat from rental to condominium status is only allowed when the flat is sold to the sitting tenant within a period of seven years

cities) are strictly restricted. At the same time, landlords who wish to increase rent revenues on their property can switch between different regulations. Put differently, German rent regulations give rise to four different gaps, the closure of which enables rental yields to be increased:

1. *New tenancy gap*: Whereas rent increases to sitting tenants are limited to a level of 20% within three years, rent for new tenants is usually agreed upon on the basis of free market rents. Theoretically, individual German states (*Bundesländer*) can limit the rents for new tenancies to a level of 10% above the price for comparable flats (*Vergleichsmiete*), yet in practice the underlying legislation is difficult to handle and is rarely applied. Thus, whenever demand exceeds supply on the housing market, new tenancies are preferable from the point of view of a landlord. The problem, however, is that tenancies are secured, so it is difficult to convince tenants to move and give way to a new tenant.

2. *Modernisation gap*: In contrast to the slow opportunities for price increases in normal sitting tenancies, modernisation activities enable landlords to push up prices very rapidly. As modernisation activities are usually costly, the opportunity to add 8% (until 2018, 11%) of the cost to the rent enables rent increases that go far beyond the opportunities provided even by new tenancies. Moreover, the likelihood that tenants can fend off modernisation is fairly limited and has been reduced over the last decades through court decisions and legislative changes. Modernisation trumps numerous tenant protections. It is, therefore, the 'silver bullet' for rent increases, and in many cases, modernisation fees are even demanded at a level that is way above market rents to motivate tenants to cancel their tenancy, opening up opportunities for new tenancies (which are then agreed upon at a level way above the existing rent, but below modernisation fees).

3. *Tenure gap*: Comparable opportunities are provided by the conversion of rental homes into privately owned homes. Here, the only legal precondition to be met is a so-called 'declaration of separateness' (*Abge-schlossenheitsbescheinigung*), which is a technical certification that the units to be converted are sufficiently separate and self-contained, alongside a division plan for the whole house.[14] Most declarations of separateness are not filed because of acquisitions of rental homes for owner-occupation, but rather in the context of buy-to-let schemes in which rental buildings containing several flats owned by one property owner are divided into smaller sections that are bought by several owners. Despite this reality, once a unit is converted, owners of individual flats can claim a need for personal use, which then justifies a

dismissal of the rental agreement by the owner. The conversion of rental homes into privately owned homes, therefore, undermines the otherwise highly protected security of tenure and enables the vacating of renters from apartments and achieving new tenancies at higher rates. Many municipalities see it as a 'fire accelerant of gentrification' (to quote Klaus Müller, a former Mayor of Berlin) and have tried to restrict the opportunities for claiming personal use, or have completely banned the conversion in certain areas.

4. *Deregulation gap*: This is the fourth gap that has opened up by the way social housing has been recommodified and turned into (less regulated) private rental housing. As already described, subsidies for social housing were considerable in the past, and this resulted in rent caps and occupational obligations for a large stock. The point here, however, is that the status of social housing is a time-limited affair in Germany that is lost once the duration of past subsidies has expired. When this happens, social housing is treated no differently from the rest of the market, and all past regulations are annulled. In Berlin, to give one example, this has been the case for more than 250 000 flats in the preceding 15 years and, as a consequence, rents have increased above the level of the local rent price index (*Vergleichsmiete*). As these increases are made possible only through deregulation, I suggest calling this a 'deregulation gap' (see also Novy 1990).

Summarising the shape of the German housing system in a nutshell, one can speak of a system of bastions and trenches built into a market economy. This system has historically emerged as a consequence of both extreme states of emergency after the two World Wars, but also as an outcome of the political and regional fragmentation of German politics. Based on the constitutional principles of a social market economy, it reflects a moving consensus between the interests of landlords and tenants. In this system, both interests count to some degree and a compromise between them is institutionalised along a set of parameters that are frequently modified, but have never been given up in total. As the whole system rests on the functioning of markets that are seen as superior to state allocation, gentrification is enabled, but at the same time restricted.

The Russian Experience

Russia has been in 'crisis' for as long as anyone can remember. . . . For well over a century, the country has been looking for a viable social order combining economic dynamism with political legitimacy. Various modernization strategies have been pursued by different political regimes, yet Russia remains a resolute laggard in competitive terms compared to the advanced western industrial societies.

(Sakwa 2011, pp. vii–viii)

What is true for Russia in general, is particularly true for Russia's housing sector. Not only have Russian housing conditions been rather dire compared with their Western counterparts throughout most of the last 100 years, but moreover the country has also seen two rounds of radical ruptures in which the existing orders have been overthrown and replaced by completely different modes of ownership, production and distribution of housing. Both of these 'revolutions', the communist coup d'etat of 1917 and the (re)introduction of capitalism after the dissolution of the Soviet Union in 1991, have widely failed to solve the enduring housing problems and to bring about some kind of normality, yet both have set essential conditions without which the shape of the current housing dynamics in Russia cannot be understood.

Housing in the Soviet Union

When the Bolsheviks took over in 1917, housing was to play a central role in their ambition to create a new society. The 1920s in particular saw radical debates in architecture and planning and considerable experimentation with new, communal forms of housing and habitation (see French and Hamilton 1979; Bater 1980; Andrusz 1987; French 1995). Arguably, the most important outcome of the revolution with regard to housing was the nationalisation of most land and housing and its transfer into municipal and state ownership. Under Soviet rule, personal ownership of housing was possible but very rare in urban areas. Land was considered separately from buildings and could not be owned by private individuals at all. As a consequence, in the Soviet system, housing was a reward given to those seen as deserving by the state. Housing allocation was an administrative procedure, not a market issue.

The norm for housing distribution in the Soviet Union was nine square metres per person in the 1920s[15] (but rarely achieved in most urban centres). While in theory this mathematically calculated and bureaucratically distributed norm would have provided at least equal access to living space (albeit in poor conditions), in practice there were considerable inequalities between different social groups. Specialists in various cultural and social spheres, the high ranks of the Communist Party, and other 'exemplary' people were often rewarded with an apartment, thus establishing a hierarchy of consumption with privileges for certain groups of workers, while not so privileged residents had to wait for a long time in the queue to get an appointment.

Ex-criminals were excluded from the system altogether (Hjödestrand 2009; Vihavainen 2009). While housing distribution in the Soviet Union can, thus, in general be described as equality in poverty, there were also advantages for a privileged few, as well as considerable disadvantages for those at the margins of society.[16]

What the Bolsheviks had inherited in 1917 was an extreme housing shortage, overcrowding and an underdeveloped system of planning and urban self-government. Under the Tsarist regime, housing conditions were poor and there was a severe shortage of housing in the rapidly industrialising urban centres. One of the first measures taken by the *Bolsheviki* to deal with the disastrous housing situation (which was first put into practice in Petrograd[17] in 1918) was the redistribution of housing (*zhilishnyi/kvartirnyi peredel*),[18] i.e. the municipalisation and redistribution of bourgeois houses to the working classes. In practice, this meant taking away the luxurious homes of the rich and giving the individual rooms to workers for free. The major outcome of this operation was the emergence of a new housing type – the communal apartment (*kommunalka*) – as a normal tenure. In a communal flat, a number of families (usually between two and seven, depending on the size of the apartment) shared a flat, with each family having one room and sharing the use of the kitchen, the bathroom and the hallway. While in practice there were also experiences with this form of communal living in many urban slums in the western parts of Europe at this time, what this policy did was to establish *kommunalki* as the main type of habitation, especially in the historical centres. This was particularly the case for St Petersburg. Being Russia's former capital, its centre is full of spacious palaces and mansions built for the Tsarist aristocracy, resulting in redistribution in extreme numbers. In the 1920s and 1930s, communal apartments became the main housing type in Soviet cities; thus, for example, in 1935 between 84% (in new housing) and 90% (in old housing) of families in Leningrad lived in communal apartments. Only about half of them had a separate room, the rest shared rooms with other people and slept on the kitchen floor, in hallways or in other similar areas (see Vihavainen 2009, p. 43).[19]

At least to some degree, these housing conditions were also an outcome of a policy of rapid industrialisation and urbanisation under Stalin. Priorities were set to foster industrial development, while state investment in housing and other parts of the 'consumption fund' came second. Together with the enormous destruction caused by World War II, this led to an increasingly intensive housing crisis. It was only after Stalin's death that Krushchev, the new leader of the Communist Party, set new goals for housing policy. In a famous speech delivered in 1955, Krushchev said that architects of the neo-classical buildings favoured in the Stalin era 'build monuments to themselves', rather than useful space for society (Zavisca 2012, pp. 29–30). He demanded that they considerably lower building costs and drastically increase supply. Simultaneously, Krushchev aimed to 'de-Stalinise' the living conditions of the average Soviet citizen by providing them a basis for a normal life, with some degree of stability and security. There were two major outcomes of this political turnaround. The first one was an enormous acceleration of housing production, achieved by the introduction of standardised technologies. The urban housing stock nearly doubled between 1955 and 1970 (Zavisca 2012, p. 29), and prefabricated housing became typical for Soviet cities, as did living in a separate apartment. While housing conditions, thus, improved considerably in general, it also needs to be noted that the quality of many of these buildings was not particularly high and the planning and building of infrastructures and high-quality living environments remained a big problem. Moreover, the focus on new construction led to an intricate relationship with the existing building stock, which was rather neglected and under-maintained. The second major feature of Krushchev's policy was the establishment of new norms for housing distribution. Under Krushchev, the norm for the distribution of housing was kept at nine square metres of living space, yet with the goal to provide a separate apartment for each family until 1970. With this, fundamental changes in the housing culture were set in place. Owning a separate apartment for family-oriented housing – instead of communal apartments – became the norm and a primary goal to be achieved in the average Soviet citizen's lifetime. As Zavisca (2012, p. 41) described, 'having had a separate apartment – along with a car and a dacha – is a sign of having lived well'. In general, these lines of housing policy were also followed by Krushchev's successors, yet the aspiration of providing a separate home for each family was never achieved. At the end of the Soviet Union, 20% of families were still on the waiting list for housing and another 25% hoped to join it (Kosareva and Struyk 1993).

Regardless of this failure on their own terms, Soviet housing policies had considerable effects on the Russian housing culture and the expectations of most Russians. There are different aspects worth mentioning in this context. First, housing provision and housing conditions had been rather poor for most households in Russia for a long time. There has always been a considerable shortage of housing and, thus, it is not uncommon to have three generations of a family sharing a small apartment or even a room. Second, notwithstanding these conditions, and in sharp contrast to most capitalist countries, an outcome of the Soviet system of housing distribution was that Soviet citizens in general came to see housing as an entitlement provided by the state and earned by hard work, political function or good conduct – rather than something acquired in a market. According to Zavisca (2012, p. 6):

In the Soviet Union, housing was a right for citizens and a reward for socialist labour. Everyone who worked (and those incapable of working) was entitled to shelter, but work more valued by the regime was rewarded with better housing.

Third, tenant rights were very strong in the Soviet Union and rents were kept stable and low. Eviction for non-payment was practically impossible. At the same time, the equipment standards provided were rather low (as were building quality and maintenance). Improvements to the individual housing situation were, thus, often done in the form of do-it-yourself investments, such as equipping balconies with windows to gain additional living space, repairing leaks in the roof, etc.

The state distribution of housing space, the low rents and the individual investments made by many residents successively led to a culture in which tenants developed a feeling of quasi-ownership, in which they treated their state-owned apartments as if they were their own.[20]

While it is self-evident that this history has contributed to a number of issues for today's housing in Russia, the implications of more than 70 years of communist rule in the field of housing are rather difficult to summarise. Any attempt to do so would, however, need to deal with a number of contradictions. The most important ones seem to be the emergence of a deep-seated sense of entitlement and quasi-ownership, a chronic housing shortage, poor living conditions, notorious underinvestment in the existing stock and resulting maintenance issues, the legacy of *kommunalki* and a generally very low degree of sociospatial segregation, together with massive privileges for certain groups (for the latter see Szelenyi 1983). As will be shown in this section, these contradictions still shape the housing situation in Russia today.

From Shock Therapy to Failing Markets[21]

After the dissolution of the Soviet Union in 1991, the new Russian government led by Boris Yeltsin set a vast reform in place. The major goal of this reform was not so much to improve the housing situation, but to abolish the established system of state production and distribution of housing and to create a housing market from scratch. Thereby, the overall reform strategy put forward by President Yeltsin's team was clearly based on a Chicago School style of economic thinking. Russia's shift from socialism to capitalism was a paradigmatic example of a shock therapy, and the reforms implemented primarily focused on the privatisation of a maximum of state-owned assets, price liberalisation and state downscaling. The aim was not only to radically reform the system in general, but also to achieve this goal immediately (actually within about a year[22]). Tellingly, the reformers 'viewed themselves as a "kamikaze" government whose task it was to carry out an irreversible transformation of the state-run system in the shortest possible time' (Cook 2007, pp. 60f.).

In the housing sector, this policy was implemented under the auspices of a United States Agency for International Development (USAID) financed Housing Market Reform Project (see Zavisca 2012, pp. 49f.) run by the Washington DC based Urban Institute. Paralleling the neoliberal orientation of the overall reform policy of Yeltsin's government, the major orientations recommended by the American advisers clearly adapted the typical laissez-faire economics of that time, as they were advocated by the Washington Consensus. Their foundational pillars can be described as a quasi-religious belief in the supremacy of unchecked markets, a dogmatic view on unregulated price formation and an apotheosis of private property and minimal state intervention. Homeownership was central to this project, both because it was seen as a precondition for the establishment of markets and a value in itself, but also because it was expected to give Russians a stake in the market reform (thus, bringing the UK's Margaret Thatcher's vision of a 'property owning democracy' to its extreme). In order to enable private ownership to contribute to free price formation, the state needed to withdraw from all sorts of price regulation. Housing subsidies were to be abolished or transformed into means-tested individual allowances. Mortgage markets were also central to the project, as a provision of credit was fundamental to enabling long-term investments. With the goal of channelling a maximum of capital into the system, these were designed as 'a Russian copy of the American secondary mortgage market system'[23] (Mints 2000).

The major practical steps taken on this basis were: (i) the transfer of state and enterprise owned housing to the municipalities, (ii) the privatisation of this housing to its residents, (iii) the withdrawal of subsidies and their replacement with residents' payments, (iv) the establishment of a federal housing mortgage lending agency and (v) the introduction of means-tested assistance to eligible households. In a nutshell, the reforms aimed to move housing off the state's budget, make it a tradable commodity, and to shift the responsibility for its production and maintenance to the consumers.

The reform started with the Law on Housing Privatisation in July 1991, which was amended to the Law on the Fundamentals of Housing Policy in December 1992. The raison d'etre of the new law was that it gave tenants of state and municipal owned or departmental housing (i.e. housing owned by companies) the statutory right to claim individual ownership for the unit they inhabited. In a way, this copied the Right to Buy policy introduced by Thatcher's government in the UK in 1981, while at the same time bringing it to its extreme, as privatisation would be at no cost and comprise nearly the total housing stock. The main features of this privatisation were regulated as follows (for more details see Struyk and Kosareva 1993):

- Tenants could decide by themselves whether they wanted to privatise or not. Yet, only tenants officially registered as the occupants could purchase their flat.

- Everyone was entitled to privatisation, but only once. Under-aged children could thereby become co-owners of the apartments in which they lived with their families.

- After a slow start during which the purchase was regulated by vouchers given in relation to the average size and quality of housing in the respective city, privatisation was free of charge.

- During the transition period (the end of which was not specified in the law and has been postponed several times up to now), the old system of housing provision remained in effect for non-privatised units.

- Households had no obligation to decide immediately, but could exercise their right to privatisation at a later point in time. Thereby, households that had been given a unit in the time following the start of privatisation had the right to purchase it under the same conditions as those who already inhabited an apartment at that point in time.

Very soon it became clear that the privatisation strategy was met by unforeseen difficulties. Privatisation had a quick start and residents rushed to claim their apartments, especially those living in well-located and high-quality apartments and senior citizens (who couldn't hope for a second opportunity to do so later). However, after this demand was saturated, privatisation considerably slowed down in the second half of the 1990s. Thus, in 2000, only half of the eligible units were privatised and many residents did not seem to wish to privatise their apartment in the foreseeable future.

There are a number of reasons that explain this reluctance. Citizens with a place on the municipal waiting list often waited to be assigned to a better apartment hoping to privatise this one later on. Others feared the maintenance, insurance and operation costs that might be increased due to private ownership in a time of economic insecurity and crisis. A third group of people distrusted the idea of owning a home in general and were hesitant to take responsibility for what could potentially be a burden in the future. Changes in personal circumstances could also play a role and lead to the need to reconcile property issues with structures of inheritance and other complicated intra-family consideration ((Zavisca 2012 gives a number of telling examples of this). Given, the economic crisis following the transformation and widespread insecurity in all matters of personal and professional circumstances, this type of calculation was common, and privatisation proceeded very slowly. Altogether, this resulted in probably the longest privatisation process of the world, which even now – more than 25 years after its start – has not yet finished.

In addition to the much weaker than expected appeal this policy had for the Russian people, the giveaway privatisation implemented also had a number of flaws on a more general level.

First, by privatising the flats to sitting tenants, it perpetuated inequalities inherited by the Soviet system of housing provision. Thus, tenants who had a favourable position in the former system had a higher chance to have well-located, high-quality assets than those who were disadvantaged in the Soviet Union (for example, former inmates, workers in less prestigious professions or people who had just not been on the waiting list long enough, or were too young to claim their own apartment). Thus, there was a problem with matters of justice, as assets were unevenly distributed. This problem has also been recognised by those who constructed the law (see Kosareva and Struyk 1993) but has been justified as the necessary price for creating a market.

Second, the law created a paradoxical situation in which a household's income and its ownership of housing assets were largely decoupled. It, thus, enabled the emergence of asset-rich but income-poor citizens – as well as others who had a high disposable income but no assets. In this context, it became absolutely common to find unemployed or retired Russians who could hardly meet their everyday needs living in spacious flats in the most prestigious housing stock, as well as families who were doing fairly well, but could hardly afford a tiny flat in an unfavourable location. The reformers had foreseen this problem too, but largely relied on the emergence of markets, which they believed would set into place an effective mechanism of housing allocation. As will be discussed later in this chapter, this strategy was built on sand, and the outcome is a pattern of housing allocation that often seems fairly irrational in the eyes of Western observers.

Third, while the flats were privatised, the land and the buildings remained municipal property. Housing associations in which the owners of a building would take common responsibility for maintenance and operation costs would only be established in 2005 and still struggle with difficulties. The general upkeep of the building is, therefore, to a large degree provided by the municipality, which charge the owners *kvartplata* (literally 'flat pay') at administratively defined rates. Even with privatised condominium buildings, the state has a considerable influence, as the municipality usually acts simultaneously as (i) the co-owner of the common property in the building (because it owns the not yet privatised apartments), (ii) the owner of the housing management and utilities companies and (iii) a public administration, responsible for planning issues, such as budget allocation and social housing provision (see Puzanov 2012, p. 224). Private ownership of apartments and public/cooperative ownership of the buildings, thus, do not easily match in Russia, and in buildings next to one another built prior to 1991, the state is still at least a co-owner. Moreover, in practice, the state and municipalities continue to finance major repairs,[24] and a large part of the utility costs. As a consequence of this situation, the actual composition of monetary flows is often confusing.

Fourth, while the state was quick to privatise flats and also implemented the basics for private or cooperative ownership of multifamily buildings after a while, it was way more reserved with privatising land. While the 1993 Constitution of the Russian Federation introduced private land ownership in Russia in general, it did not detail the conditions for it. It was only in 2001 that a new Land Code of the Russian Federation was enacted and clear procedures for the privatisation of land were defined. Nevertheless, most of the land in Russia is still owned by different state authorities and hardly sold. Thus, to give but one example, in St Petersburg 82.7% of the urban land is state property, while 13.5% of the land is owned by corporate bodies, and only 3.8% belong to individuals (data from the Municipal Committee for Property Relations [from http://www.kzr.spb.ru/info-land.asp, no longer available]). These figures are fairly constant over the last decade. As a general rule, state-owned land is leased to investors rather than sold. As a consequence, the implementation of all sorts of investment projects that do not take place on land sold from the minor share of privately held lots is totally dependent on governmental bids. Instead of controlling capital-driven urban development with planning guidelines and zoning regulations, the state, thus, still holds enormous power as the major landowner in most cities, and investors must follow leasing and purchasing procedures with public authorities that are often described as frustrating and risky for private investors (see Limonov 2011).

Fifth, and as a consequence of all these circumstances, privatisation led to an enormous degree of fragmentation with regard to ownership, income and asset strategies. Within one building, it was not uncommon to find families occupying the flat they privatised, cheek by jowl with others who bought the flat they inhabit using a mortgage or cash payments, and again with others who rented the flat from a private owner who lives elsewhere, as well as residents who rejected privatising their flat and were still municipal tenants, plus those that received their social flat because they were eligible for social housing. This situation is even more complicated in the case of *kommunalki*, where individual rooms or even parts of rooms were privatised. The coexistence of owner-occupied, private rental and state-owned apartments in one building, or even one apartment, is thus rather the rule than the exception and 'scattered' property structures have become a common feature of Russian cities. To make matters more complicated, the tenure of one flat can even change back and forth within a short period of time. Thus, for example, owner-occupiers can decide to move to their datcha over the summer and let their flat to tourists. Inheritors can sell a flat they privatised at no cost, in order to achieve a form of heritage that can more easily be spilt among different heirs. A flat can also change its status from rental to owner-occupied when the owner needs it for personal use or for their family and then return to renting it out at a later point, and so on and so forth. As these different uses entail different orientations towards either the value of the flat as a consumption good or as asset that is bought, sold and rented out in the market, this state of affairs often leads to conflicting and hard to oversee valorisation strategies.

Price liberalisation, the second pillar of the housing reform, proceeded with an even slower speed than privatisation. In Soviet times, housing costs were heavily subsidised by the state and were in fact seen as a kind of a second pay packet, distributed outside the workplace, regardless of status or income. Rents were fixed in 1928 and remained at a very low level of 2–4% of disposable income throughout the Soviet era. A consequence of this history is a widespread sense of entitlement to affordable housing that still characterises the mindset of many Russians on this subject and coexists with neoliberal and conservative narratives. The outcome of this has been a somewhat contradictory situation with regard to the legitimisation of housing reforms. Simultaneous to the privatisation and shock therapy, Yeltsin's government enshrined a 'right to housing . . . free of charge or for affordable pay' as an entitlement to be guaranteed by the state in the 1993 constitution (see Zavisca 2008, p. 372). Given a situation of enduring economic crisis and very low average incomes, raising housing costs was politically risky, as it confronted deep-seated beliefs about the rights and entitlements held by the general public and also affected wide groups of people in a very problematic way. As a consequence, until now, *kvartplata*, which is defined by the state, is only increased rather hesitantly. Although the average expenditures for rent and utilities had increased by 230% by 2000, they still made up only 6.5% of the total household income on average in the early 2000s (UNUCE 2004, p. 25), and this figure has hardly changed since. Only half of the utility costs were paid for by the residents, whereas the rest was paid for by the municipalities. As for rents and utility payments today, both are very affordable (if measured by Western standards) in the municipally owned and the (outright) owner-occupation sector. In addition, there are numerous exemptions and discounts for different groups. Non-payment is common, but in practice the risk of eviction is low. This situation makes housing and utility subsidies a remaining burden for municipal budgets, 'accounting on average, for about 40% of municipal budgets . . . crowding out expenditures on health, education, and other municipal obligations' (Cook 2007, p. 177).[25] Thus, although Russia did move towards a liberalisation of housing prices, it did so with the brakes on, and the outcome was 'a hodgepodge of residual tenancy rights, subsidies, means-tested benefits, and partial rent payments . . . governed effectively neither by the state nor by market principles' (Cook 2007, p. 143).

Dysfunctional property structures and impeded price liberalisation have not been the only barriers to the neoliberal project in the housing sector. Until now, a functioning mortgage market offering long-term, low-interest mortgages is also largely absent. Only about a quarter of transactions for house sales are done using mortgages (AMHL 2015, p. 22), and the conditions for taking a mortgage are not favourable. As of 2015, mortgage rates were around 12–13% and include a down payment of 20%.[26] As a consequence, most housing deals are done in cash. With the strengthening of a state-owned Agency for Housing Mortgage Lending (AHML) and a number of subsidy schemes have been introduced by the Putin government during the last years, and against a background of economic stabilisation, taking a mortgage has become common: nevertheless, only a minor share of Russian households qualify for obtaining a mortgage. According to Shomina (2014), a study performed by the Federal Ministry of Regional Development estimated that only 24% of all Russian households in total were capable of purchasing their own apartment, be it in on the basis of cash or with the help of a mortgage. Although the number of home loans has more than doubled since 2006 and the overall mortgage lending volume has grown by more than five times in the last 10 years (AHML 2015, p. 15), the mortgage market is overall still underdeveloped. In 2013, the total volume of mortgages accounted for only 4% of the national gross domestic product (http://tass.ru/en/economy/727331) – which is piteous compared with countries like Germany, the UK, or the USA, where the ratio is around 50–60%.

The reason for this underdevelopment of the housing finance sector can be found in the context of economic crisis, inflation and hyperinflation, domestic currency depreciation and political insecurity – as has been characteristic for the situation in Russia for most of the time since the beginning of the transformation. In a climate of enduring insecurity, long-term investment strategies are a risky endeavour, and this is reflected in widespread reluctance to engage in long-term lending and borrowing.[27]

With the presidency of Vladimir Putin, housing became a regular topic in public speeches, and the unsatisfactory state of affairs was discussed at the highest levels of government. Thus, in his second term as a president, Putin declared openly:

One of the sharpest socio-economic problems in today's Russia is the large difference in incomes. . . . Providing the low-income population with housing is very important in relation to this problem. But let us be frank: mortgages remain inaccessible to many. Many people have no prospects for solving their housing problems.

(cited in Zavisca 2012, p. 63)

In several speeches, Putin described the provision of a 'civilised market' that enables potential buyers to realise their aspirations as an economic and moral imperative for the state. Consequently, under the Putin government, considerable energy was expended to consolidate and improve the housing situation in Russia. 'Affordable and comfortable housing for the citizens of the Russian Federation' was, thus, declared one of four national priority projects, and serious new subsidy programmes were launched. Thereby, the lion's share of money and effort went into supporting households to purchase new apartments and the construction industry to build more apartments. This included subsidised credit and tax incentives for builders and manufacturers, but, more importantly, new mortgage subsidies for young families and 'maternity capital' (a bonus for childbearing that can, among other purposes, be used for purchasing a new home). In addition, new regulations were set in place that aimed to stabilise the mortgage lending sector. In summary, the new national project was majorly focused on supporting the construction of new owner-occupied housing. While this policy proved to be fairly successful with regard to the overall volume of new constructions, it did less well with improving affordability, as prices increased along with new demand. Moreover, even with subsidies, acquiring a mortgage remains out of reach for most Russians. Sponsoring ownership, thus, hardly helps to mitigate the situation when there is a lack of housing for millions of families who cannot afford to buy a house.

Much less was done for social housing and for private tenants. Here, the major milestone was the adoption of a new housing code on 1 March 2005 (which replaced the outdated Soviet Housing Code from 1983). For the first time, it defined criteria for eligibility for social housing, thus, effectively reducing the queue on the municipal waiting list. It also ruled that social tenants would pay only for utilities, not rent, and made evictions for non-payment possible (but still hard to implement in practice). While the new code clarified the division of power between different state levels and assigned the provision and management of social housing to the municipalities, it did only a little to establish the financial preconditions for fulfilling this task. In practice, most municipalities are overburdened, and there is a serious lack of financial capacity to meet even the mandatory tasks (see Kollegova 2011).

Restricted State Capacities and Opportunity Planning

Inefficiency, lack of experience and an overload of bureaucratic procedures are other characteristics of the situation, and most Russians have little optimism that the state has the capacity and the interest needed to change the situation described earlier in this section. In fact, one of the most puzzling observations to many is the fact that state structures and legal procedures in Russia seem to be overly bureaucratic, rigid and authoritarian, yet at the same time often laxly applied and interacting with a shadowy world of informality, untransparent decision-making and corruption.

One of the most interesting models to tackle the apparent contradictions between formal political institutions and informal practices has been developed by the British political scientist, Richard Sakwa. Sakwa conceptualises Russia as a 'dual state' (2010, 2011) and describes the situation as follows:

> Contemporary Russian politics can be characterised as a struggle between two systems: the formal constitutional order, what we call the normative state; and a second world of informal relations, factional conflict and para-institutional political practices, termed [here] the administrative regime. . . . We argue that a dual state has emerged on the basis of the two systems, and much politics takes place in the zone between them.
>
> (Sakwa 2010, p. 185)

According to this model, the coexistence of formal democratic institutions and informal decision-making is, thus, not a contradiction, but essentially the way public life functions in Russia. Rather than mutually excluding each other, both systems remain in place at the same time, and each carries political weight. Thus, while legal institutions define the playground, the political game is determined by 'telephone law', corruption and informal networks between political and economic interests. At the same time, the formal constitutional and legal order provides opportunities for contestation. The outcome is a stalemate between the normative state and administrative practices, in which neither one nor the other can solely operate on its own logic. This results in a continuously shifting landscapes of power with dynamics that are rather untransparent and bound on networks of interpersonal relations and individual advantages.

It needs to be emphasised that the coexistence of formal institutions and interpersonal networks of decision-making is nothing new for Russia (and, in fact, for much of the world). Tensions between formal and informal modes of governance have been in place in Russia since before the revolution and throughout the Soviet Union. What is new, is the fact that this system now works in a capitalist economic environment. In practice, capitalism and democracy do not, therefore, stand against informality, bureaucracy and corruption in Russia. Quite in contrast, 'state institutions [that] do not work' (Kononenko and Moshes 2011) and informal governance, or *Sistema* (Ledeneva 2006), can provide opportunities for business interests. Capitalism and corruption are two sides of the same coin here, which together can be either very functional or very detrimental to individual capitalists; it depends. In this sense, it has become something of a truism to talk about Russia as a 'hybrid regime' (Shevtsova 2001; Diamond 2002; Hale 2011; Petrov et al. 2014), as an 'imitation democracy' (Furman 2008) or as 'crony capitalism' (Sharafutdinova 2010). Sharafutdinova characterises the main features of this situation as follows:

> When a close connection between state officials and economic elites dominates policy making, traditional cronyism becomes 'crony capitalism' . . . a distinct institutional order characterized by the domination of informal elite groups. In such a system selected economic elites receive preferential treatment and privileges, making support from the state rather than market force a crucial factor for maintaining and accruing wealth. It is no secret that the years of postcommunist transformation in Russia and many other states undergoing similar transformations resulted in the emergence of such a system.
>
> (Sharafutdinova 2010, pp. 2–3)

With these logics in place, both economic and political elites must maintain and aim for control over bureaucratic decision-making procedures in order to obtain and keep privileges. The consequence is the maintenance of strong bureaucratic procedures, coupled with an inclination for tricks, behind-the-scenes agreements, nepotism and corruption.

There are numerous consequences of this nexus for urban development in Russia that cannot be discussed in detail in this chapter. Instead, I will only point towards two issues that have relevance for the issue of gentrification in St Petersburg and that are discussed in Chapter 6.

The first of these is the fact that urban planning in Russia is still, to a considerable degree, marked by its provenance from Soviet practices (which were marked by a strong emphasis on technical issues, a high degree of centralisation and an expert-centered style of decision-making, see French 1995) and embedded into a strictly hierarchical system of government that has even strengthened under the Putin government (and is usually referred to as the 'power vertical' in Russian). This origin has facilitated a rather functional planning style in which much emphasis is placed on physical developments and functional zoning definitions to be performed under the strict auspices of different hierarchical layers of government, with only a small amount of public participation (Golubchikov 2004; Trumbull 2012). In this environment, integrated and procedural approaches towards urban planning issues have a difficult road. Besides standing in opposition to inherited planning cultures, including numerous stakeholders necessarily weakens the power of individual state bureaucrats to establish and maintain preferential relationships with individual businesses. The outcome is not only, as the American geographer Nathaniel Trumbull put it, that 'theory and reality have had, as a rule, little to do with each one another in the context of urban planning in post-socialist St Petersburg and likely in Russia at large' (Trumbull 2012, p. 388), but there is a general reluctance towards working on 'wicked issues'. This is particularly problematic because the decentralisation of public responsibilities to the municipalities, which has taken place since the 1990s, was hardly accompanied by adequate funding and tax collection structures. So, cities in Russia, as a rule of thumb, lack many of the resources they need to fulfil their tasks.

Second, the intermingling of public administration careers with business interest has facilitated a change towards a rather specific kind of entrepreneurialism, in which cities are ready to roll out the red carpet to private developers for often large-scale development projects. This entrepreneurialism has, however, not replaced the existing formal and informal structures of planning and decision-making. Weak enforcement capacities, corruption and the general intertwining of political and economic interest have, therefore, rather led to flexible strategies, in which existing regulations are kept in place but suspended, misinterpreted or bent on a case-by-case basis. Scholars of post-socialist urbanisation have termed this situation 'opportunity planning', i.e. 'a shift in planning from controlling urban development to enabling piecemeal development that – notably – benefits urban governments financially' (Taşan-Kok 2006, p. 62).

Conclusion: Gentrification in a Dysfunctional Market

Quite obviously, the expectation of the shock therapists of the early 1990s that the 'invisible hand of the market' would by itself create a self-regulating housing market once property rights were established and prices liberalised proved to be overly naive. Neither privatisation, nor price liberalisation, nor the development of housing finance worked out in the way planned. Also, Russia did not transform into a classical capitalist democracy, but into something of a hybrid in which the economy follows market rules in general, but bureaucratic and informal practices remain powerful. The major consequences of this situation are threefold:

First, in most capitalist societies, housing markets are built on a number of typical preconditions like private ownership, availability of capital, price formation through supply and demand and an allocation of housing opportunities as a function of disposable incomes. In Russia, none of these characteristics works well. Housing opportunities and residential mobility are not predominantly determined by markets, but to a larger degree by non-market forms of housing distribution like inheritance, prevailing administrative practices of price regulation for housing and utility costs, and highly differentiated patterns of housing rights for different social groups. In addition, markets for housing finance only function for a minority of Russian households, and there is a notorious lack of opportunity to realise personal demand due to an inefficient system of capital provision and housing production. The failure to transplant USA-style neoliberalism into a post-Soviet context, has led to severely impaired markets in which housing wealth is largely a frozen asset, long-term financing is not available for the majority of Russians and mobility is impeded. Under these conditions:

> ... what has emerged is not a housing market, but a regime of property without markets in which housing is privately owned but incompletely commodified. In this peculiar order, housing status bears little relationship to wages. Post-Soviet property allocations result mainly from state redistribution and family reciprocity, not market situation.
>
> (Zavisca 2012, pp. 87–88)

In this situation, the connection between status, income, assets and spending power, which is so fundamental for the working of segregation in Western housing markets, is considerably weakened, to say the least.

Second, the course of housing reform has led to a splintering of housing rights within the same building, often even within a single flat. While private and public housing have not yet formed fully distinct sectors and there are traces of social housing in next to all segments of the housing stock (see Puzanov 2012), with regard to the question of how housing costs are determined and how vulnerable households are protected against eviction and displacement, four groups of residents can be distinguished.

The first group consists of owners who have privatised their apartment and live in it (so to say 'genuine' owner-occupiers), or their heirs. As discussed earlier in this section, certain groups of low-income residents (pensioners, residents of high quality apartments, etc.) had a higher incentive to privatise and, as a consequence, a considerable percentage of owner-occupiers have rather low-incomes and little capacity to invest money into the upkeep of their property. At the same time, the housing costs paid by this group are also minimal. At the present, owners pay a housing unit tax, plus a small fee for maintenance and utility costs. In practice, owners enjoy all the usual advantages that come with property rights and low housing costs, but more often than not they are incapable of carrying the burden of the maintenance and upgrading of their properties. In many respects, low-income owner-occupiers are trapped within the property they use, as moving out would increase their housing costs in a very considerable way and the chances of finding suitable alternatives in the immediate environment are often minimal.

The second group consists of owner-occupiers who bought their apartment on the market, either using cash or with a mortgage. Like the first group, these residents enjoy all the typical ownership rights and only pay a rather low *kvartplata* to the municipality and/or the homeowners association they are a part of. Yet, in addition to this, they might easily be burdened with the costs of the loan taken to acquire the flat. Under the conditions of the Russian mortgage market (see the section 'From Shock Therapy to Failing Markets'), these can be quite considerable. Within this group, owners who financed their property using a mortgage in a foreign currency are particularly vulnerable, as these are highly affected by currency devaluation, as has taken place several times in recent Russian history, and can lead to dissatisfaction as shown in Figure 3.1. In general, owners who purchased their flat have a rather stable income. However, most property deals are signed in the newly built housing stock, rather than in existing buildings, so the share of this category is quite minimal in the historical centres.

FIGURE 3.1 Homeowner protesting mortgage increases due to currency inflation (photograph: Matthias Bernt).

The third group of residents is composed of both tenants who refused to privatise their apartments and, thus, still have a status as a municipal tenant and social tenants who have been given a municipally owned apartment due to their low income. In Russian legislation, the units inhabited by both of these groups are termed 'housing stock under social rental agreement'. Tenants who fall under this category still enjoy quite a wide range of rights that can be traced back to the quasi-ownership status of tenants in the Soviet Union (see the section Housing in the Soviet Union). Thus, in stark contrast to the private rental market, rental contracts in this part of the housing stock are signed for an indefinite term. Moreover, rents are very low, and tenants can swap their apartment with other municipal tenants, bequeath their contracts or even sublet their flat.

Eligibility criteria, housing costs and the social composition of the residents of the municipal housing stock fall into three categories, due to the fact that this stock, to a large extent, consists of the 'leftovers from nonprivatized state and municipal housing' (Puzanov 2012, p. 228) and the fact that restrictions on joining the municipal waiting lists for housing were only introduced in 2005. Thus, this part of the housing stock includes sitting tenants who didn't want to privatise their flat, tenants who couldn't privatise but were already on the waiting list before 2005, and 'genuine' social tenants in the Western understanding of the term. Only for the latter do income limits apply. In addition, there are a number of other categories of people (war veterans, teachers, physically impaired residents, etc.) who can also be put on the waiting list for municipal housing.

Rents in the housing stock under social rental agreement are set annually by the municipality and consist of 'pure rent' and a fee for maintenance and repair services, plus utility costs. Residents who meet the current eligibility criteria are excluded from paying the pure rent. However, the pure rent covers only a rather symbolic portion of the total cost; the major share (around 60–70%) is made up by the utility costs. Furthermore, certain groups of people can apply for housing allowances, for which the key parameters are set by the municipalities and the regions.

The fourth group of residents, which largely falls through the cracks of public regulations, is made up of private tenants who rent their flat on the market. As Shomina (2010) has illustratively put it, this group is a sort of a 'forgotten minority' in the Russian housing system. While, private renting has, by and large, been ignored by the legislation, its share in the market is quite considerable and growing. Due to the dubious nature that this form of tenure still has in Russia, it is difficult to estimate its actual size and share of the market. Russian experts rate it at 350 000–400 000 households in St Petersburg alone (http://www.nevastroyka.ru/info/eslisdayete-kvartiru-v-arendu-nado-platit-nalog), i.e. around 15% of the total housing stock.

As a general rule, private rental contracts are limited for a certain (and usually very short) period. Rents are freely agreed between the landlord and the tenant, and tenants have no legal protection against eviction. In reality, moreover, many tenants are not given contracts at all, as it is quite common that owners want to avoid paying taxes on their rental income. Often, flats are let with furniture and equipment, so that tenants are fairly restricted with regard to the arrangement of their personal environment. The rents paid in this sector are very high compared with average incomes and are comparable to those paid in Western Europe. In summary, whereas all the other groups of residents described in this section enjoy certain rights, private tenants are exposed to the full force of an unregulated market environment.

While somewhat similar compositions of tenure can also be found in many Western cities, what makes the Russian situation different is that all the described forms of tenure can be found in very close proximity (usually in the same house, and sometimes even in the same apartment), and that the described categories are intermingled. Thus, it is not uncommon to own a privatised apartment in one city and rent in another. Also, poor owners often move to family members (or in summer times to their datcha) and rent out their apartment to acquire some additional income that can help them to make ends meet. Sometimes, people can be co-owners of a flat privatised by their parents but pay high prices on the private market because they cannot live in the same apartment for a number of reasons. Conversely, poor tenants who have been eligible for a municipal social flat can become asset-rich overnight if they inherit a valuable flat privatised by relatives. Summing up, the relation of different tenures is rather complex in Russia. This is all the more important as both the rights connected to a particular status and the actual housing costs related to it differ in an extreme way.

Table 3.5 shows the approximate housing costs to be paid in different tenures for a two-bedroom flat of approximately 80 square metres in the centre of St Petersburg.

As the table demonstrates, the differences between the prices paid according to different tenures are very considerable. Residents who bought their flat or have to rent it on the private market pay up to ten times the amount paid by outright owners (mostly residents who claimed ownership through privatisation) or municipal tenants. Even if one takes into account that the quality of housing provided on the market is usually higher, immense differences remain.

What share do these different kinds of tenure have in St Petersburg? In general, housing statistics are very insufficient in Russia and many changes (e.g. renting without registration) can hardly be grasped using

TABLE 3.5 Estimated housing costs for a two-bedroom flat in the centre of St Petersburg (2017)

Tenure	Approximate monthly costs
Owner-occupied with mortgage*	80 000–90 000 roubles
Private rental	70 000–90 000 roubles
Outright owner	10 000 roubles
Municipal rental	6000–8000 roubles

* monthly payments on the basis of a mortgage taken for an apartment at the price of 7 million roubles, with 10% down payment, 12% interest rate and 5% repayment rate.
Sources: emls.ru; www.kvartplata.info; own estimates based on interviews

administrative data. In a rather unique study conducted on the basis of data from the Municipal Housing Management Service, Kornev (2004) studied the share of different tenures in 43 houses in the central districts of St Petersburg. Within this sample, the share of individually owned flats in the buildings observed was between 7% and 80%, of private rentals between less than 5% and 50% and of communal apartments between close to 0% and 60%; between next to 0% and up to more than 30% of the single-family flats in the individual houses were not yet privatised (ibid., p. 10). In a nutshell, the conditions between each of these 43 residential buildings varied a lot and no spatial pattern that could explain these differences could be found.

The implications of this hodgepodge for gentrification are complex. Thus, whereas in private rental apartments, economic displacement can take place immediately and tenants can easily be evicted once the owner finds this appropriate, the situation is more complicated in all the other categories. As already discussed, individually owned flats have usually been acquired through highly differentiated ways (including privatisation at no cost, inheritance and purchases using cash or credit) and, as a consequence, the likelihood that owners are willing and able to cash in on potentially higher values of their properties is varied. Moreover, an important share of the apartments is still owned by the state – mostly because the inhabitants refused to claim ownership – and cannot be sold on the private market. Instead, they are rented out at very low cost. On top of this come communal apartments, in which all these situations can be found in different rooms. Thus gaps between actually achieved rents and potential rents exist – but the opportunities for exploiting them are limited to a minor share of the apartments in each house, and there is not even a geographic logic to this.

Could this deadlock be overcome by state action? Much Western literature has ascribed a prime role to neoliberal states in supporting and facilitating gentrification (see Smith 1996; Hackworth and Smith 2001; Porter and Shaw 2009). Portraying the state as an active supporter of gentrification would, however, be a gross misinterpretation in Russia. Here, the boundaries between private and public interests are floating, and the implementation of all kinds of policies (no matter their colour) is often inefficient. This not only leads to a considerable blurring of interests between businesses and state actors, but most of all to flexible strategies and piecemeal approaches in which policy orientations and project implementation are decided upon on a case-by-case basis. As for now, the transformation of the housing sector has resulted in a hybrid between inherited Soviet traditions, failed neoliberal shock strategies, informal and crony government structures and discreet steps towards correcting and consolidating some the of the most problematic issues arising from this.

In this context, gentrification is seriously impeded by (i) the remaining lack of connection between tenure, income and social status and (ii) the splintering of different forms of tenure on a geographic micro-scale. Both have by now severely restricted the effective functioning of a 'normal' capitalist housing market and made gentrification difficult to achieve. At the same time, the housing shortage persists and the distribution of housing opportunities is highly uneven and dysfunctional. The system remains in crisis.

State Intervention in Housing: Setting the Parameters for Gentrification

In this chapter, I have aimed to show how the provision of housing has developed as an area of conflict between capital accumulation and social rights in three very different societies over the course of roughly one hundred years. In none of the three societies has housing provision and consumption been solely a matter of market allocation.

As gentrification has been restricted in different ways at different times and places, some systematisation of the main parameters along which the state can influence housing production and consumption can be useful. Roughly, policy options towards the regulation of housing can be located along the stages that occur in the process of providing housing, including, as shown in Figure 3.2) state interventions on financing, development, construction, allocation and maintenance, in the production side, and policies towards the purchasing power of households, on the consumption side (see Lundquist 1992).

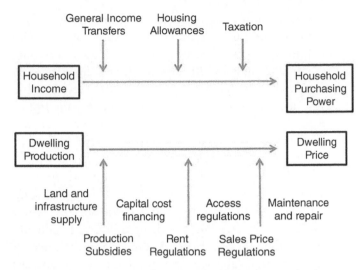

General Income   Housing
   Transfers   Allowances   Taxation

Household Income → Household Purchasing Power

Dwelling Production → Dwelling Price

Land and infrastructure supply   Capital cost financing   Access regulations   Maintenance and repair

Production Subsidies   Rent Regulations   Sales Price Regulations

FIGURE 3.2 Diagram of state interventions in the housing sector.

States can intervene both in the production and the consumption of housing. Both forms of intervention can be achieved using numerous instruments that can be combined in manifold ways. At the same time, the choice of instruments is not contingent at all. In contrast, it usually operates in close relation to the existing welfare state model and the specific housing market at a given point in time. Intervention in housing provision is, moreover, a 'wobbly pillar of the welfare state' (Torgersen 1987; Malpass 2003). More often than not, it is politically contested and vulnerable to change. Russia and the UK are illustrative examples of this.

The picture is even more complicated, as different forms of intervention can mean different things in different contexts. Not only are there numerous forms of state intervention in the housing markets, but each of the instruments mentioned in Figure 3.2 can be organised in varied ways. Moreover, different instruments can also be combined in quite different ways, meaning that state intervention can take fairly varied routes.

Production subsidies, to give but one example, can thus take various forms, and their effectiveness depends on the composition of housing providers, on the price and accessibility of land, on the conditions on the credit markets, and on many other factors. Moreover, subsidising the production of housing can be very selective with regard to building types and locations, thus fostering uneven geographic patterns.

The same is the case for forms of tenure. Tenure composition is crucial to the matter of gentrification because tenures define the formal rights of residents in their position as owners, tenants or social housing dwellers and set up the rules of the game. Decision-making on tenure forms, thus, 'defines the bargaining room for seller and buyer, landlord and tenant, together with economic support for different types of housing in terms of subsidisation, financial security or tax relief' (Bengtsson 2015, p. 680). This, however, does not allow for simple comparison, as the same tenures in formal terms in fact vary in content to a great degree in different countries. While renting from a private landlord, for example, is synonymous with a lack of security (at least since the deregulation of private renting in the late 1980s) and best to be avoided in the UK, this is not at all the case in Germany, where renting is the most dominant and highly regulated tenure. The meaning and stratification effects that go hand in hand with ownership in the UK, or in Germany, are completely different from the Russian 'hyper-ownership society', where residents were 'given' their flats for free and valorising this asset is difficult. Whereas council housing in the UK was widely decoupled from the market for a long time, in Germany there has always been subsidisation of commercial development projects, linked with temporary rental and occupational obligations. In Russia, in contrast, social housing more or less refers to the leftovers from privatisation. In summary, the social relations underlying a specific tenure are highly diverse.

Why is this relevant for gentrification? As I have aimed to demonstrate in the previous chapters, the political regulation of housing not only sets the conditions for housing provision in general, but also defines the parameters for gentrification. Reinvestment depends on the exploitation of differentials that are defined by the different valorisation opportunities offered or denied by housing policies for different parts of the housing stock. Gentrification, one could say, works its way around existing regulations. As these regulations not only differ between different countries, but also between different tenures, and change over time, we end up with a fairly complex picture. In a nutshell, the picture looks as follows for the three countries studied.

In the UK, gentrification has for a long time been synonymous with a shift from private renting to owner-occupation. National policies have supported this transformation with tax benefits and subsidy schemes that over-proportionally have benefited middle- and upper-class households, thus fuelling gentrification and the displacement of low-income tenants. At the same time, council housing (as a state-controlled and decommodified form of habitation) worked as a massive bulwark against gentrification. Both private rentals and council housing enclosed potentials for capital accumulation that could be exploited once the tenure was changed. The consequences were the emergence of (classical) value gaps, resting on the difference between the value of a tenanted home and a vacated home which could be sold for owner-occupation, and the gap between the status of council homes as a non-commodity and their actual market value if sold on the market. With deregulation, globalisation and house price inflation, this picture has changed since around the turn of the millennium, and by now more and more households that would have traditionally been termed 'gentrifiers' are excluded from buying a home and are relegated to the private rental market. With this, gentrification is now also taking place within the private rental sector, whereas more and more groups are excluded from participating in the sector by buying a house. Geographically, this has been reflected in the simultaneous ripple effects expanding gentrification ever further outwards, state-led gentrification of social housing estates and the subsequent super-gentrification of already gentrified areas, together producing a landscape of intensive and interlinked gentrification dynamics.

This history is a far cry from the German experience, which is marked by a dominance of renting and a complex system of regulations towards this form of tenure. For most of the last 100 years, housing distribution in Germany was controlled by the state, and it is only since the late 1960s that market allocation has gained ground. Despite an unmistakable neoliberalisation, housing reforms have always proceeded with the brakes on, and the specific features of the German state have resulted in a system of dugouts from which both landlords and tenants can exercise some powers. This, however, does not prevent gentrification – but it modifies its modus operandi. German rent laws, thus, allow for three different gaps (change from a sitting to a new tenancy, modernisation and transformation of rental flats to individual ownership) in the rental sector in which yields can be considerably increased. In addition, the temporarily limited status of social housing enables enormous profit gains when the duration of subsidy programmes come to an end and the building is returned to the 'normal' market. While these gaps enable gentrification, they also impede it and modify its course to some degree. As in the UK, this framework has produced its own geography in which locational effects are counteracted and sometimes neutralised by the differentiated protection status of the households and flats situated at a particular location. As most protections are limited with regard to durability and scope, this also leads to a dynamic landscape in which the enabling/disabling of gentrification fluctuates geographically.

In Russia, the situation is again different, and there are enormous differences between returns that can be gained in different types of tenure. The problem here is that the lion's share of the stock is privately owned but barely commodified. The same goes for maintenance and renovation, which are, by and large, state-financed. Housing allocation is thus not only a market process, but a matter of family reproduction and state provision. While gentrification is, thus, theoretically possible, it is de facto made very difficult by a fragmentation of property rights. The outcome is an extreme splintering of the process in which different households live in the closest spatial proximity, but worlds apart with regard to the conditions of their housing consumption.

Summarising this bewildering diversity of conditions, it becomes clear that gentrification is possible in all three contexts – but only under parameters that differ to a bewildering degree. What potential and capitalised ground rent are caused by, the actual value and the ways in which ground rent can be captured vary greatly between the UK, Germany and Russia.

In this sense, this chapter has identified 12 commodification gaps that have depressed the potential ground rent to an actually achievable level. Both by the way they have limited ground rent capture and by the loopholes they provide, these gaps have effectively determined the actual operation of gentrification in the UK, Germany and Russia at different times. Table 3.6 provides a brief overview of these gaps.

As is easily seen, the conditions for the operation of gentrification have varied greatly: both between different national contexts, and also within one

TABLE 3.6 Commodification gaps in the UK, Germany and Russia

Country	Time	Gap	Explanation
UK	1960s–1980s	Value gap	Difference between the tenanted value of a flat and the value of a vacated home sold for owner-occupation (see Hamnett and Randolph 1984)
	Since 1981	Privatisation gap	Gap between the status of council homes as a non-commodity and the actual value of the respective flats, if sold on the market (made possible through the Right to Buy policy)
	Since 1988	Rental deregulation gap	Gap between potential and capitalised ground rent in private rentals in the UK emerging through the abolition of rent controls in 1988
	Since 2000s	Inflationary gap	Gap between potential ground rent and capitalised ground rent resulting from the inflation of house prices
	Since 2000s	Derivative gap	Gap between capitalised ground rent and potential rent resulting from the securitisation of land titles

TABLE 3.6 (Continued)

Country	Time	Gap	Explanation
Germany	Since 1960s	New tenancy gap	Difference between the capitalised ground rent for a flat rented out with a new contract and the rent that can be demanded from a sitting tenant
	Since 1974	Modernisation gap	Gap between the rent for an unrefurbished flat and a modernised flat resulting from the legal opportunity to increase existing rents by 11% of the modernisation costs after renovation
	Since 1980s	Tenure conversion gap	Gap resulting from the conversion of rental flats into privately owned individual homes. This weakens the security of tenure and leads to either the sale of the flat for owner-occupation (→ value gap), or re-letting (→ new tenancy gap)
	Since 1990s	Social housing deregulations gap	Gap between market rent and subsidised rent in a social housing units effectuated the expiration of the social housing status after subsidisation period
Russia	Since 1991	Tenure conversion gap	Difference between the rent capitalised for an owner-occupied home when it was privatised by the sitting tenant and the potential rent that can be gained when renting out or selling the unit in the market
	Since 1991	*Kommunalka* gap	Difference between the rent capitalised for a *kommunalka* apartment and the rent that can be gained when this tenure is dissolved
	Since 1990s	Redevelopment gap	Gap between the rent capitalised for an existing building and the rent that can be potentially gained when the building is demolished and a new structure is built on the plot; demands the lifting or bypassing of preservationist regulations

nation at different times and within particular locations. The differences can only be regarded as contingent when the point of reference is an abstract market model uninterested in history and agency.

When these are factors included, the picture gets more complex. Throughout this chapter, I have demonstrated how changes in the setup of the parameters for gentrification have been a result of political struggles, in the form of both class mobilisations and passive revolutions (set into motion by ruling elites to keep their power). Each of these struggles resulted in new 'condensations' (Poulantzas 2000 [1978]), i.e. in different shapes and forms of 'social rights' (Marshall 1950) being built into the housing market through rent regulations, mortgage provision, social housing and/or systems of tenure. As these factors and their combination have differed between Russia, Germany, and the UK, and also within these countries over time, so have the respective housing systems and the parameters that have contributed to the emergence of commodification gaps.

Notes

1. In Labour-led London, nine-tenths of all post-war housing was provided by local authorities and over half of this was constructed by the Labour controlled London County Council alone between 1946 and 1961. From the 1960s to the 1980s, the borough councils embarked on further massive house building programmes. As a result of these programmes, close to one-third of all London households lived in local authority flats in 1981, whereby the number was considerably higher in long-term Labour boroughs such as Tower Hamlets (82%) and Southwark (65%) (see Watt and Minton 2016, p. 209).

2. Although this dual pattern seems to replicate the respective ideological preoccupations of the two parties governing Britain since the early twentieth century, two words of caution are necessary to prevent oversimplification. First, emphasis on, and support for, council housing have cooled down within the Labour Party over the course of time and, as early as the 1970s, the idea that buying a house was 'a basic and natural desire' (DoE 1977, p. 50) was supported by the Labour Party. Second, until 1979, notwithstanding ideological reservations, Conservative governments only hesitantly intervened into the existing system and did not stop the expansion of council housing when they were in power. The reason for this was that only local authorities were seen as having the organisational capacities necessary to deal with the housing shortage, especially after World War II, so that playing the 'numbers game' and demonstrating power to deliver large numbers of new council homes became part of the political competition. Moreover, building more council housing was also seen as being instrumental for slum clearance, so that inner city 'regeneration' was

often implemented in the form of bulldozing Victorian terraces and replacing them with modernist estates run by the state.

3. The concept of a property owning democracy claims that private property ownership promotes the necessary independence of mind and social stability for the responsible exercise of political power (Jackson 2012). In the British context, it has been used by the Conservatives since the early 1900s and seen as an alternative to state and collective ownership and a means to protect individual independence both against state tyranny and despotism and against the concentration of power in the hands of corporate capitalists. Margaret Thatcher revived this world-view in her first speech as the new party leader and made it a centrepiece of her policy when she became prime minister.

4. Rachman was a notoriously ruthless landlord and speculator in London in the 1950s and 1960s. Today, Rachmanism is used as a synonym for the exploitation and intimidation of tenants by unscrupulous landlords (for more information on Rachman, see Davis 2001).

5. In the UK, council tenants have the statutory right to be balloted on any privatisation proposals. This has led to the defeat of around a quarter of all privatisation proposals in total. Thus, for example, tenants stopped stock transfer and demolitions in Birmingham in April 2002, with over two-thirds of the tenants voting against these changes.

6. While the history of housing policies under German state socialism is a topic worth a discussion in its own right, I refrain from including this issue here. The reason for this is the fact that socialist housing policies were largely discredited after 1990 and hardly had any influence on the shaping of housing policies in the reunified Germany. Policies introduced in the territory of the former GDR after reunification have by and large been set up as a copy of the Western standard model and only been slightly modified where this was inevitable. I will pick up on these modifications in Chapter 5 on gentrification in Prenzlauer Berg, where necessary.

7. Karl-Christian Führer (1995, p. 15) has correctly pointed towards the negative bias included in the term *Wohnungszwangswirtschaft* that is usually used in Germany to mark this period. It would be as correct to speak of the antidote *Mieterschutz* (tenants' protection policies) to characterise the regulations put in place.

8. Of these, rent arrears was the most important point as given the typical structure of rentals in Germany where a multistorey building with 20–30 households is usually held by one landlord, claiming a need for personal use could not affect the majority of tenants. It deserves to be noted that 'disorderly conduct' included unmarried sexual encounters and, in particular, homosexual relations until the 1950s.

9. This is also true for the much-acclaimed reformist architecture of this time (e.g. the 'horseshoe settlement' in Berlin-Britz). In the majority of cases, these socialist experiments were too expensive for most workers.

10.   In the long run, this resulted in a perverse system in which cost rents came to be much higher than market rents, thus justifying enormous public subsidies and rent increases in the social housing stock (see Holm et al. 2016)

11.   The number of social housing units in Berlin, to give but one example, decreased from 370 400 (1993) to 114 000 (2016) in two decades (Gewos 2016).

12.   In the 1980s, non-profit housing companies lost much of their previously good reputation when a corruption scandal shook the trade union owned flagship of the sector and led to its bankruptcy.

13.   Red is associated with both the Social Democrats and the ex-Communist Left Party (DIE LINKE) in Germany, while green stands for the Green Party.

14.   Here, building morphology plays a role. Contrasting with much of the English-speaking world, the most common building form in inner cities are multistorey residential homes in which several families live on one floor but in separate flats. Traditionally, a complete building was owned by one landlord who would rent out the flats in the house. In the 1970s, this changed and the property could be divided horizontally. This has given way to the sale of individual flats and has led to a situation where a building has up to 40 different owners, including both owner-occupiers and buy-to-let landlords.

15.   Interestingly, the norm focused on total living space instead of the number of rooms, which did not take into consideration the different shapes of apartments.

16.   In communist ideology, this was justified by the idea that equal access to consumption goods could only be provided in a fully realised communism, which was yet to be achieved, while developed socialist societies would still need to distribute these gods according to the work that is performed in the interest of the society.

17.   St Petersburg was renamed Petrograd from 1914 to 1924 and then given the name Leningrad from 1924 to 1991.

18.   The operation was also called 'filling up' (*uplotnenie*).

19.   For interesting insights into the realities of communal living see https://russlang.as.cornell.edu/komm.

20.   The same does not really apply to public and semi-public spaces like yards, stairways or pavements, which are often terribly neglected and not taken care of in Russia.

21.   Parts of this section have been published in Bernt 2016b.

22.   Thus, in January 1992, 90% of wholesale and retail prices were released from state control overnight – which resulted in a devastating economic crisis, hyperinflation and the devaluation of wages and savings for most Russians.

23.   It is both striking and telling that Russian reformers, thus chose the one variant of a mortgage system that later on proved to be most vulnerable in the course of the subprime crisis in 2008.

24. Major renovations are, thus, mostly financed on the basis of a state fund for 'capital repairs' in which the owners of the housing units pay only 5% of the renovation costs (see Puzanov 2012, p. 225).

25. It was only after the political and economic situation had stabilised that the new Putin government dared to take the next step towards price liberalisation. The so-called 'Gref Plan' called for transferring 100% of utility costs to the residents within two years and introducing means-tested housing assistance for the poor. It was immediately softened in intragovernmental discussion and its realisation postponed to 2010. Even this version caused strong resistance by the Duma, and in the spring of 2002, the introduction of pilot reform programmes caused large demonstrations and a public boycott of housing and utility payments. In this situation, the government managed to gain a majority in the parliament only by making numerous concessions. Housing privileges for teachers, war veterans, teachers, and other groups were preserved; evictions were made more difficult; the obligations for poor households were halved, and the requirement for 100% payment dropped (Cook 2007, p. 178).

26. Until around 2000, the conditions were even more difficult. Then, average loans were for six months to two years on interest rates between 17% and 30%, and they were usually denominated in US$ 'Such loans were only affordable to the very small fringe at the top end of the qualified professionals group' (Minto 2000).

27. This structural difficulty is complemented by other problems. Thus, in a survey conducted in 2003 (cited in Mints 2004), Russian banks identified the lack of effective foreclosure and eviction as the main barrier to more mortgage lending, followed by difficulties in attracting long-term funds and the handling of unreported incomes (which reflects widespread tax avoidance and fraud and makes it difficult to establish reliable information on the credit history of potential borrowers).

CHAPTER 4

Barnsbury: Gentrification and the Policies of Tenure

Barnsbury is a residential neighbourhood in the North London Borough of Islington, which emerged around 1820 as a middle-class suburb. For many Londoners, Barnsbury is synonymous with classical gentrification and:

> . . . it would be difficult to find a more classical or attractive gentri-
> fied area in Inner London than this. . . . The highly desirable houses
> and cottages, many of them beautifully, almost perfectly, renovated, line
> spotless, ordered and cared-for streets; the area would appear by all to
> be beyond further improvement – on a mild spring day the ambience
> of luxurious idyllic.
>
> (Butler and Robson 2003, pp. 51 52)

What makes Barnsbury interesting for the purpose of this study is the fact that the area has undergone gentrification over a very long time, at least since the late 1960s. The history of urban upgrading dates back half a century, so the impact of changing economic and political environments can be studied over an unusually long period.

The Making of Early Gentrification

The Barnsbury area[1] can be seen as one the birthplaces of gentrification in the UK, and many of the early studies on gentrification in London (Chambers 1974; Hamnett 1973, 1976; Hamnett and Williams 1979; Hamnett and Randolph 1984; Williams 1976, 1978; Power 1972, 1973; Pitt 1977) had the then novel changes in this part of Islington as their focus. What happened here, as well as in parts of Kensington, Paddington, northern Lambeth,

The Commodification Gap: Gentrification and Public Policy in London, Berlin and St Petersburg,
First Edition. Matthias Bernt.
© 2022 John Wiley & Sons Ltd. Published 2022 by John Wiley & Sons Ltd.

Hammersmith and Camden was at that time a sensational transformation of Inner London and the Berlin-born German émigré and University College London professor Ruth Glass was the first to coin the term gentrification to describe it (see Chapter 1).

Back in the 1950s and 1960s, Barnsbury was, however, anything but elegant and expensive. Quite the contrary, post-World War II suburbanisation and the subsequent outmigration and disinvestment led to considerable abandonment, and the steady outmigration of younger and more skilled English families left a population of unskilled, often foreign–born, young families, and an ageing English-born population behind. Barnsbury increasingly became an area for those who were either too old or too entrenched to move away, too poor to buy a home in the suburbs, or not able to get on the list for council housing elsewhere (see Lees et al. 2008, pp. 10–11). What had been planned as a middle-class suburb, increasingly became one of the most notorious inner city slum areas of London. Housing stress was severe. In 1961, nearly two-thirds of Barnsbury's households lived in shared accommodation and a study conducted for the Borough of Islington in 1968 found that of 168 households interviewed, 127 had no access to a bath, 138 shared a toilet, 15 had no kitchen sink, and 25 were living in overcrowded conditions (Lees et al. 2008, p. 12).

This changed rapidly in the course of the late 1960s to early 1970s. While it is a fact that the changes linked to gentrification became an issue for many areas in Inner London, making Barnsbury anything but a unique case, they were most articulate in this neighbourhood. Here, the increase in the percentage of economically active males in the upper segments 1, 2, 3 and 4 of the UK census social classes was most pronounced with an upwards movement from 3.3% in 1961 to 15.8% in 1971 (Hamnett and Williams 1979, p. 12), whereby the starting point was lowest among all gentrifying areas in London. In other words, the transformation from shabby and modest to elegant and expensive described by Ruth Glass in the quote cited in the first chapter, covered the widest social distance and at the same time went on at the quickest speed.

Class change in Barnsbury was accompanied by a striking tenurial transformation. Over a period of 50 years, the area saw two waves of fundamental transformations with regard to percentage of households by tenure, as Table 4.1 demonstrates.

Thus, the 20 years from 1961 to 1981 saw the demise of a housing market dominated by private rentals and its shift towards a market split between council housing and owner-occupation. As most council housing was built in the clearly distinguishable southern parts of the ward, the changes in the more rapidly gentrifying areas of terraces and detached houses in the north were

TABLE 4.1 Percentage of households by tenure in Barnsbury (Butler and Lees 2006, p. 473; UK Census 2011, own calculations). Note that there were significant boundary changes in the census between 1971–1981 and 1991–2001

Tenure (%)	1961	1971	1981	1991	2001	2011
Owner-occupier	7.0	14.7	19.0	28.9	34.0	33.0
Council and social rent	15.0	18.7	56.0	48.7	47.8	38.5
Private rent	75.0	65.2	13.9	10.7	16.6	28.5

even more pronounced. Since 1981, the share of council and social rented housing has continuously decreased, in favour of both more owner-occupation and (later) more private renting. In a nutshell, in Barnsbury, as in most of Inner London, neighbourhood change has not only been a matter of changing occupational and class compositions, but it has also been accompanied by a remarkable tenurial transformation (see also Hamnett 1991; Watt 2001).

The first wave of this transformation, the change from private renting as the dominant form of tenure to owner-occupation, has been described by Hamnett and Williams (1979) as follows:

> In 1966 at least half of the properties in Lonsdale Square in the Barnsbury area of Islington were let to furnished tenants and other properties were let as unfurnished. Earlier than that almost all were tenanted. By 1975 the Square was distinctively middle-class, only six of approximately fifty properties in the Square were still rented and some of these were part owner occupied.
>
> (Hamnett and Williams 1979, p. 481)

Social class change and the transformation of the housing stock from rentals to owner-occupation were closely connected. Whereas, on the surface, the changes in the social composition of the residents of Barnsbury taking place in the late 1960s and early 1970s can be read as an outcome of the immigration of young and well-educated 'invaders' (Bugler 1968) caused by new lifestyles and a new taste for inner city housing, a closer analysis reveals that this process was guided by a switch of valorisation strategies based on different tenures. Gentrification was actively 'produced' and real estate speculators and 'residential asset strippers' (Hamnett and Randolph 1988, p. 6) had a key role in this production.

The major theoretical argument about this 'flat break-up market' (and the political economy of tenurial transformation it was based upon) has been made by Chris Hamnett and William Randolph (1984, 1988). Within the gentrification literature it has become known as the 'value gap' thesis. In short, the value gap[2] is the gap between capitalised property value for rental and owner-occupied properties. It reflects the dual value system in residential property that has emerged in Britain since the 1920s and describes the conditions it set for reinvestment into the existing building stock. Thereby, as an outcome of changing fiscal and financial structures (which will be discussed in more detail), the 'tenanted investment (TI) value' (i.e. the capitalised rental income achieved from a property let to a tenant) came under increasing pressure in the second half of the twentieth century. Consequently, selling property for owner-occupation became more lucrative. As owner-occupied property is exchanged on the basis of vacant possession, Hamnett and Randolph called the value of a property transformed to ownership 'vacant possession (VP) value'. The point here is that 'those landlords who sold to owner-occupiers did . . . so . . . in order to make capital gains by appropriating the difference between the capital values of their properties in the rented and owner occupied markets' (Hamnett and Randolph 1988, p. 70).

The emergence of a value gap is, thus, crucially embedded into British housing policies at a particular historical moment. It does not represent an eternally higher value of ownership vis-à-vis renting, but it reflects the higher profitability of selling for owner-occupation in specific circumstances. The value gap rested on a number of institutional and political preconditions that taken together resulted in making the sale of rentals for owner-occupation more lucrative than letting. When these changed, the value gap disappeared, as will be shown later. As the gap was based on the differences in profitability of two forms of tenure, it could effectively also be called a 'tenure gap'.

Having this in mind, the most important issues leading to the emergence of a value gap in British inner city neighbourhoods in the 1960s are explored in this chapter.

First, subsequent inflation, as it was to be experienced in the UK in the late 1960s and 1970s, diminished the value of capital invested into rental housing. Whereas in periods of low inflation, long-term tenancies and controlled rents could still be seen as safeguarding a landlord's investment, in times of high inflation, rentals tended to become a liability for the landlords. The inability of rental yields to keep pace with inflation ate into the value of capital and led to a successive loss of the profitability of being a landlord relative to other investments. In this situation, owner-occupation increasingly became an alternative. One reason for this is that property valuation for owner-occupied property is very different from the one for rental property. Whereas, in the case of rentals an investment into housing is valued on the basis of attainable rental income, what matters for owner-occupiers is capital gains. Thus, in contrast to rentals, inflationary pressures can even be welcomed because they increase the value of property. 'Indeed, house price inflation and the capital gains which flow from it form a major attraction of homeownership' (Hamnett and Randolph 1988, p. 72). Put differently, the more inflation becomes an issue, the more owner-occupation becomes advantageous to the investor.

In addition, and this brings us to the second point, both the availability and the price of a mortgage and the taxation of property played a crucial role. In this respect, UK taxation clearly worked to the advantage of owner-occupation. While the structures of taxation have changed over time, all British governments since World War II have supported homeownership (see Chapter 3) and have provided incentives for owner-occupation. Whereas rental incomes were made subject to taxation based on the current rent levels received, owner-occupiers were taxed on the historic value of their property fixed at infrequent historic intervals. As a consequence, for the owner-occupier the effect of this tax tended to decrease over time, whereas it was constantly maintained for the landlord (Hamnett and Randolph 1988, p. 74). Moreover, mortgage interest tax relief, which was only abolished by the New Labour government in the late 1990s clearly subsidised acquiring properties by making payments on mortgage interest deductible from one's income for tax purposes. This made acquiring property very attractive, especially for high-income taxpayers. Finally, unlike landlords, owner-occupiers have been largely exempt from paying capital gains tax since it was introduced in 1965. In sum, British tax policies worked in favour of owner-occupation from the 1960s (Holmans 1987), and the effect of this was (i) owning was made more attractive than renting for households able to make it with a mortgage and (ii) the buying power of potential homeowners was expanded. Tax laws, thus, privileged owner-occupation compared with letting, and this resulted in a long-term downturn of the private rental market.

A fairly similar effect was brought about by the expansion and flexibilisation of the mortgage markets. Given that house purchase for most households requires obtaining a mortgage, the decision of mortgage lending institutions as to what types of properties and which types of borrowers they are willing to lend to is a crucial determinant for the viability of owner-occupation. In this regard, two major changes in mortgage lending policies happening in the course of the 1960s were crucial for the realisation of the value gap. First, building societies successively increased the percentage of property valuations advanced over time to 90% and lengthened the repayment period to 25–30 years. This made acquiring properties more manageable for more of the middle-classes, thus, increasing the demand. Second, as Peter Williams has shown (1976), lending policies changed from 'redlining' to active support for regeneration projects in inner city areas. Until the mid-1960s, the main financial institutions were rather reluctant to lend on the type of properties to be found in these inner areas, as well as to the type of blue-collar and immigrant applicants who generally came forward there. Thus, until the early 1960s 'there existed a financial void in Islington' (Williams 1976, p. 74) and larger building societies, banks and insurance companies hardly made any loans in this area. As the 1960s progressed, this changed successively and, by 1972, all but one building society had granted mortgages in Islington (Williams 1976, p. 77). Again, the motives behind that change were very much connected to a changing political environment. Williams notes a number of political decisions, which taken together, helped to change the mortgage policies of financial institutions, among them being: £100 million given to building societies under the 1959 House Purchase and Housing Act with the aim of increasing owner-occupation in older properties; the 1967 Leasehold Reform Act, which enabled tenants with long leases to acquire the freehold at advantageous prices; and the Option Mortgage Scheme of 1967, intended to increase the opportunities for low-income households to purchase a house, as well as considerable improvement grants (see also Hamnett 1973) for the rehabilitation of older housing. Taken together, all these policies made acquiring property less risky for the borrower in regard to both the area and the type of housing. 'Thus, as confidence increased, so did the means by which change was enabled' (Williams 1976, p. 78).

Whereas tax policy changes and modifications in mortgage lending were clearly instrumental in supporting owner-occupation and house purchases, the situation was more contradictory with regard to rents and rental incomes. In general, the rent controls introduced in 1915 had only been subject to minor changes since and the Conservative 1957 Rent Act (which had removed the statutory restrictions on rents and introduced main steps towards rent decontrol) was scrapped when a new Labour Government came into power in 1964. Thereby, the strengthening of rent controls by Labour was largely a reaction to the poor housing conditions in Inner London and 'major problems with landlords who made life a nightmare for some tenants' (Hamnett 2003, p. 133). Most important in this context was the Rachman Scandal[3] which showed connections between a particularly ruthless landlord and the Conservative establishment and led to an increased public awareness of the violence and intimidation practices some private landlords had used against their tenants. The Rachman scandal forced the government to set up the Milner Holland Committee to investigate the problems of housing in London (see Crook and Kemp 2011, p. 11), and when the report was published, it made clear that landlord abuse was a common problem in the private rented stock. Reacting to this situation, the new Labour government introduced another Rent Act in 1965 that introduced 'regulated tenancies' and 'fair rents' assessed by independent rent officers. The overall attempt of the new law was to provide a fairer power balance between landlords and tenants by restoring security of tenure, while at the same time providing regular opportunities for fair rent increases. The outcome of this was that letting flats was made even less attractive as a business model.

Summarising, the picture is a contradictory one. On the one hand, government policies in the UK have for a long time worked in support of homeownership, and this made selling homes for owner-occupation a more profitable business than letting. On the other, with a short-lived exception between 1957 and 1964, rents were controlled and tenants were protected against the perils of the market. This resulted in a growing contradiction between a large rental stock waiting to be transformed into owner-occupation, and the fact that this stock was still inhabited by tenants whose rents were controlled and who were unwilling or unable to buy the places they lived in. The only solution to this problem was to get rid of the tenants and sell the vacated flats for owner-occupation.

This resulted in an active 'making' of gentrification, which for the case of Barnsbury has been described by the then activist and now London School of Economics (LSE) Professor Anne Power (1973, n.p.)

> The 'London Property Letter' of February 1970, privately circulated to estate agents and property interests, recommended Barnsbury as a 'chicken ripe for plucking' ... Speculators were moving in on the calculation that a house full of tenants could be bought for £2,000, emptied for between £2,500 and £5,000 over a year, and resold unconverted for £12,000 to £20,000. In the most conservative estimate, speculators were making over 200% profit a year.

What has been termed as a value gap by Hamnett and Randolph is, thus, in many respects rather a tenure gap which was worth £9000–15 000 a year. What made this gap emerge was the support, and respective disadvantage, of different forms of capital accumulation in the housing sector by successive British governments. While this policy environment produced the conditions for making the break-up of rental housing profitable, irrespective of the actual location, a number of factors contributed to making this a particularly interesting option in Barnsbury. In other words, a value gap induced by housing and taxation policies coincided with a rent gap caused by conditions produced at the local scale.

First, as Hamnett (1973) has pointed out, improvement grants provided by the national government since 1969, with the aim of assisting the regeneration of the existing housing stock, were mainly used for property before stock conversion and sale for owner-occupation. 'Plucking the chicken' was, thus, helped by public money – yet this assistance was locally restricted and there was a strong geographic concentration in the use of these grants in gentrifying areas. Anne Power (1972), thus, reported that in 1971, 56% of all Islington's improvement grants went to the wards of Barnsbury and St Peter's. Improvement grants, one could say, thus, worked as a signal lamp marking the entry lane for gentrification.

Second, gentrification has been supported by 'pioneers' from within the neighbourhood. As Ferris (1972) has reported, early gentrification in Barnsbury was accompanied by serious conflicts between middle-class in-migrants who had formed a Barnsbury Association and both more working-class residents and more leftist supporters organised in the Barnsbury Action Group. Thereby, the Barnsbury Association managed to form a housing association, which bought a terrace of houses on the north side of Barnsbury Street in 1971. Interestingly, the costs of rehabilitation for these houses proved to be higher than anticipated and, as a consequence, most people on the council housing list were effectively excluded from renting there (Ferris 1972, p. 92). Over the years, the Barnsbury Housing Association acquired additional properties and even engaged in new constructions, and it now owns 241 flats in Barnsbury.

Thirdly, large parts of Barnsbury were designated as Conservation Areas and a traffic scheme was set in place by the Islington Borough Council in the 1970s. The importance of a designation of a neighbourhood as a Conservation Area mainly lies in protection. A London-based planner expressed this nicely in an interview: 'With Conservation Areas, the important thing is not what changes, but also what is not allowed to change and this supports the kind of neighbourhood qualities searched for by gentrifiers' (Interview, Alan Mace, 16 June 2014). As the designation of a Conservation Area considerably uplifts the cultural status of this area, it supports demand from educated consumer groups, thus increasing property values and helping to channel investments into the building stock. In this respect, it can be argued that 'conservation is a legally constituted form of profit-making' (Lees 1994, p. 214). A similar function can be attributed to traffic schemes, which assisted in reducing noise and pollution, thus enhancing an 'urban village' atmosphere (see Ferris 1972.

Summing up, what made gentrification in Barnsbury so compelling in the 1960s and 1970s, was the concurrence of policies enacted on the national scale (which worked in favour of displacing poor tenants and selling their homes for owner-occupation) with local incentives channelling the interest of investors towards this neighbourhood. The outcome was a middle-class invasion, accompanied by a change of tenure.

The Right to Buy: Pouring Fuel on the Fire

While gentrification in Barnsbury had a slow start in the 1960s and speeded up enormously in the early 1970s, by the mid-1970s it had considerably slowed down. There are different reasons for this. First, as an outcome of both a changing financial environment and changing subsidy frameworks, property values stopped rising, thereby making it impossible for developers to stay liquid by remortgaging existing properties (Pitt 1977). Second, as tenants became more militant and more organised, emptying properties by removing the existing tenants became more difficult. This was especially problematic as interest rates went up dramatically, thereby increasing the opportunity costs on owning tenanted properties. Third, also as a consequence of the oil crisis and increased interest rates, a mortgage famine severely depressed the market. Taken together, these circumstances made the exploitation of the value gap more risky and less profitable, and numerous companies withdrew from the market.

At the same time, partly due to political pressure from tenants and partly because local authorities were being urged by the central government to municipalise and rehabilitate old houses, the Borough of Islington become particularly active in acquiring properties:

> After years of disinterest, Islington Council embarked on the largest municipalisation program the country has ever seen; nearly 4,000 properties have been bought for rehabilitation by Islington Council in the last four years.
>
> (Pitt 1977, p. 11)

Municipalisation programmes were, however, not unique to Islington, but implemented by numerous local authorities across Britain. Watt (2001, p. 118), thus, reports neighbouring Camden had even acquired 12 600 housing units, representing more than one-third of the total stock. In any case, the acquisition of properties was particularly intensive in Barnsbury and accomplished both by buying up houses on the markets and by compulsory purchase orders, which were filed in accordance with redevelopment and clearance plans set up by the council. A long-time resident reported in an interview that the council had bought up complete rows of houses with the purpose of either demolishing them, redeveloping the land on which they were standing, or rehabilitating and renovating them with public funds. The consequence of this was that in the late 1970s, the Borough of Islington owned the lion's share of housing units in the area. As the borough let its properties to tenants on its waiting list, this led to a remarkable upturn in the number of council tenants, which made up 56% of the households living in Barnsbury in 1981.

When the Conservatives came into power in 1979, this tide turned. Their introduction of a Right to Buy policy (see Chapter 3), especially, had a dramatic impact on the supply-side of gentrification in Barnsbury, as it brought hundreds of properties to the market and, simultaneously, made the privatisation of council flats a source of windfall profits for the claimants. In a situation in which a share of 56% of the total housing stock was state-owned, this could have formed a massive barrier limiting gentrification; however, Right to Buy successively flattened this protective wall and fuelled gentrification through a continuous new supply of flats. Figure 4.1 shows the number of Right to Buy sales throughout London.

Pinning down the empirical details of this process is, surprisingly, rather difficult. Although, theoretically, exact data on the sale of council flats exists in the archives of Islington Council, they are in fact closed to the public. It is, therefore, only possible to work with proxy indicators, which at least help to obtain a general picture.

Altogether, 12 670 flats were sold in Islington under the Right to Buy policy between 1979 and 2018 (Department for Communities and Local Government's Local Authority Housing Statistics, https://www.gov.uk/government/statistical-data-sets/live-tables-on-social-housing-sales, table 685). Around half of these (6439) were sold in the first 20 years following the introduction of this law. As all studies on the right to buy show, the share of privatised flats was higher in well-located and high-quality stocks. Given the good location, the preferable housing stock and the early start of gentrification in Barnsbury, it is rational to assume that the percentage of the stock the council owned was

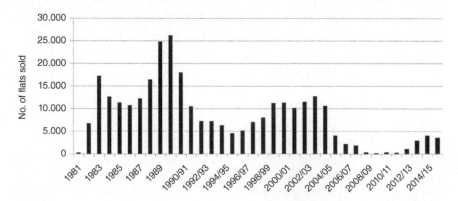

FIGURE 4.1 Right to Buy sales in London 1981–2014/2015 (Based on Housing in London, 2017).

considerably higher than the borough-wide average of one-fourth of the total council owned stock here. Given that more than half of Barnsbury households lived in council housing in 1981, it can be estimated that around 15–20% of the local council stock was sold in the years following the introduction of the right to buy.

What happened to council flats, once they were privatised? Again, there is no sufficient data available which would allow me to draw a comprehensive picture. In order to detect general trends, I have used data on homes acquired by the Council of Islington through compulsory purchase orders in the Barnsbury area for the period 1973–1976. As these orders had to be approved by the council, the actual proposals (which include the addresses of the individual properties affected) are documented in the council meeting minutes, which are accessible through public archives, and it is on this basis that I have detected the addresses I discuss here. It needs to be emphasised that these reflect only a part of the acquisition activities of the council, as many property owners preferred not to wait for a compulsory purchase order and sold voluntarily.

In a second step, I investigated whether these properties were sold, and for what price. Again, the data is insufficient here, yet it is possible to track all sales activities since 1995 through the Land Registry (https://www.gov.uk/government/collections/price-paid-data).Obviously, this data does not provide information on properties that were privatised prior to 1995 and have not been sold since then; it also does not provide any information on prices paid before 1995, and is insufficient to find out how often a property was sold prior to this date. Nevertheless, the picture that can be gained from this list is revealing.

Appendix A lists all properties bought in the Barnsbury area acquired through the use of compulsory purchase orders between 1973 and 1976 and sold after 1995. Included are 67 addresses, all of which have been bought by private owners first and then sold again by private owners at a later time. It, thus, lists properties that became council owned in the 1970s and were privatised under the Right to Buy policy later on. While it is unknown when the 67 addresses analysed were privatised exactly, for which price, and with which discount (discounts against market prices varied from 25–70% throughout the period), the list gives us an impression of the enormous profits enabled through the Right to Buy policy. Many of the dwellings included have been sold several times, and each time with considerable gains. To give one example, 183b Barnsbury Road, a three-storey Victorian tenement, has been sold four times within less than 10 years and every time at a higher price. Thereby, the most successful seller was able to achieve a profit of 80% after 13 months, and even the smallest profit made still equals a profit of more than 5% per annum.

FIGURE 4.2 Two £1.5 million ex-council houses: 163 and 165 Barnsbury Road.

A few steps down the road, at 163 Barnsbury Road (shown in Figure 4.2), waiting for 13 years made a difference of close to £1 million. At Avon House, Offord Road, half a year made a difference of £115,000 – even in a multistorey tenement with a still sizeable share of council owned flats and a high percentage of social housing tenants. Gains made from what used to be public property are, thus, enormous.

As this data demonstrates, the introduction of the right to buy generated the means for enormous profit-making. The social consequences were an acceleration of gentrification, as privatised homes came to provide a continuous supply of homes to be acquired from in-moving gentrifiers. This is all the more important, as the Right to Buy policy kicked in at the moment the London Stock Exchange was liberalised and the city experienced a massive influx of buying power. The expansion of a gentrifiable housing stock, thus, went hand in hand with a growing demand, and this resulted in the consolidation and expansion of gentrification throughout the 1980s and 1990s (see also Butler and Lees 2006).

In this process, the social relations between residents, investors, the state and land changed fundamentally.

Prior to privatisation, the plots affected were bought up by the council with the purpose of being taken out of the market and allocated to people in need of housing, for a fee. As the council didn't acquire its housing stock as a form of investment but as public infrastructure, the land and the houses built on it would become decommodified and there would be no connection between market prices and rates charged. The relation between the council and the residents would be one of a provider and recipient of public benefits and the workings of the market would only play a role as far as they would impact on the financial environment of the council's budget.

With the privatisation of council owned homes, the residents would become owner-occupiers, which made them responsible for maintaining the use value of their properties (e.g. through repairs or modernisations), but more importantly enabled them to convert their asset into cash. As the price for privatised flats was greatly discounted and at the same time housing prices increased, this meant that enormous gains were possible. Council tenants were, thus, turned into speculators, buying up property with the expectation of selling it a later time for a higher price. The borough in contrast, was left with a decreasing stock to allocate to disadvantaged households, which resulted in processes of residualisation and concentrated poverty in the council stock. Borrowing on the concept used by Hamnett and Randolph for the preceding period of gentrification in London, we can describe this whole operation as a privatisation gap or, as Paul Watt has coined it, a 'state-induced rent gap' (Watt 2009a, p. 235), i.e. as a gap between the discounted value of a flat under Right to Buy and its actual market value. The crucial point about this gap is that it rested on exploiting the capital accumulation potential piled up in the local authority housing stock through privatisation. Subjecting public land to private investment was, thus, a precondition for the gap to emerge. Yet, once this was achieved, new spaces for capital accumulation were opened.

The exploitation of this gap rested on the sale of the privatised flat, mostly for owner-occupation. In an already gentrified and well-located neighbourhood like Barnsbury, it is rational to expect that sales went over-proportionally to middle- and upper-class buyers. This reinforced the already existing gentrification pressure, and added demand for a consumption infrastructure providing goods and services to this group, thus, making the neighbourhood more attractive to other gentrifiers. By and large, both low-income residents and the borough stood aside from this dynamic and were only able to comment on the unfolding of the market, but with few opportunities remaining to intervene in the game.

The New Economy of Gentrification

In the new millennium, the economic environment of gentrification again essentially changed. Investment strategies, ownership structures and the role of gentrifiers were, once more, fundamentally modified and became increasingly unstable and contradictory. In this section, I will discuss two processes – the increased importance of property purchased as an asset (in contrast to a place to live) and buy-to-let gentrification (Paccoud 2016), which work together and make current gentrification in London distinct from previous developments.

Capital Gains Instead of Owner-Occupation

The first of these is the growing focus on housing as an asset instead of a value that gets realised through its use. The enormous inflation of house prices in London has resulted in enormous opportunities for capital gains. While housing prices have moved up in London generally at a rate far above general inflation over the last 20 years, this development is even more pronounced in Barnsbury (see Figure 4.3).

The enormous price inflation has made homeowners entrepreneurs, especially those whose mortgage has been paid off, as they can profit significantly from converting the stored-up value of their house into cash, either by selling (and moving) or remortgaging the property (without moving). This form of speculation on capital gains has considerably changed the real estate economy in Inner London over the last years and has resulted in a new form of property valorisation that is based on expected capital gains, instead of exchange values generated through private use or renting (see Whitehead 2012; Stephens and Whitehead 2013; Thomas 2014; Glucksberg 2016; Glucksberg et al. 2016; Atkinson et al. 2017).

While the boundaries between different forms of property valorisation are blurred, two mechanisms stand out in the context of gentrification.

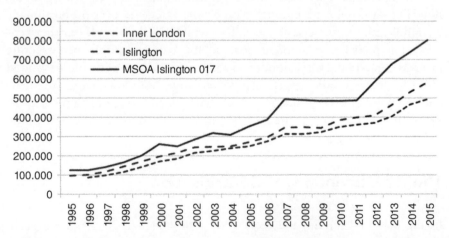

FIGURE 4.3 Median house prices (£) in Inner London, Islington and Barnsbury (MSOA Islington 017), 1995–2015 (https://data.london.gov.uk/dataset/average-house-prices).

The first of these two can be termed as 'trading down'. It includes the movement of property owners to a cheaper house or flat and them keeping the difference in value. Trading down typically happens on retirement. The second strategy is remortgaging, which provides additional regular income while using the property as collateral. The further advance gained through this second mortgage can be used to pay off existing debts or buy consumption goods (e.g. a car or a second home) or services (e.g. medical treatment, or school or university fees for the children). With both forms, property becomes not only a good for personal use, but also an asset allowing further investment or consumption.[4]

What are the consequences for areas like Barnsbury? While it is impossible to develop a reliable detailed picture of the local situation without the help of a survey-based study, some tendencies can be deducted from the general discussion about the current market situation in London.[5]

First, the price inflation excludes 'normal' gentrifiers from the property market. Between 1995 and 2012 the ratio of prices to earnings rose from 5.4 to 11.46 in Islington (https://data.london.gov.uk/dataset/ratio-house-prices-earnings-borough). With a median house price of £675 000 (2013), a household with a median annual household income of £73 820, which is typical for the already gentrified Islington 017 MSOA area (https://data.london.gov.uk/dataset/average-house-prices) would face a price to income ratio of 9 : 14. If this ratio were calculated for the median annual household for Inner London, it would be 11 : 86. Even if this household managed to afford the typical 20% deposit (£135 000) now demanded from first-time buyers by most financial institutions, it could still only borrow £283 910 (calculated using http://www.alexanderhall.co.uk/mortgages/mortgage-calculators/how-much-can-i-borrow.html) – which is not even half of the price. This has considerable consequences for the profile of buyers. The situation is especially disadvantageous for first-time buyers who can't call on the 'Bank of Mum and Dad' or other sources to afford the increased deposit and repayment cost of a mortgage. The current prices effectively restrict young middle-class buyers, thus, considerably transforming the demand-side of urban upgrading. In stark contrast to past rounds of gentrification, middle-class buyers (who are portrayed as the typical stormtroopers of gentrification in wide swaths of the literature) are effectively excluded from entering the market and in-migration into the area is becoming more and more restricted to even richer households. The outcome is 'spatially displaced demand' (Hamnett 2009), i.e. a ripple effect whereby price increases in one area push prices up elsewhere, thus leading to an expansion and generalisation of gentrification pressures in ever more parts of London.

Second, equity becomes more important, either in the form of cash or in saleable or mortgageable assets. Again, an example might help to illustrate the situation. If an individual or a household was the outright owner of a house in the Islington 017 MSOA, for the which median house price was £799 998 in 2015, they could easily trade down to a comparable house in neighbouring Holloway without even taking a mortgage and still have close to £290 000 left. This sum could then be used, for example, to purchase a second house in a less expensive area, as collateral for another mortgage, to buy a holiday flat in Spain or acquire a Rolls Royce. By now, trading down has become widespread and both conversational and journalist accounts in London are full of examples of people who bought property in a rundown area cheaply in the past, earned tremendous amounts of money when selling, and can now afford to spend their retirement time at the Costa del Sol or to send their children to expensive universities. The polarisation and sharp inflation of prices in gentrified neighbourhoods in Central and Inner London thus work to the advantage of those who already have assets and enable them to acquire more assets. This thus reinforces social inequalities, even inter-generationally, and intensifies the trend of already gentrified neighbourhoods becoming elite enclaves.

Third and connected to this, international buyers play an increasingly important role in London's real estate market (Glucksberg 2016). While London has always been a more international market than other European cities, the scale of this investment has increased dramatically in the last years. 'In 2009, it was reported that inward investment totalled £2.4 billion. In 2010, it rose to £3.7 billion . . . In 2011, that investment increased to £5.2 billion' (Heywood 2012). In 2012, Savills estimated that more than £7 billion has been invested by international buyers into prime real estate in London (Savills 2013). The scale and importance of this investment is reflected in the fact that '£5.2 billion in 2011 [is] . . . a larger sum than the whole Government investment in the Affordable Housing Programme for England in four years'. (Heywood 2012, p. 15). Several reasons for this trend have been discussed, including the desire of international investors to obtain a UK residence, a search for a place to shelter savings against economic insecurity or political troubles in the homelands, money laundering and an interest in rental yields and capital growth. The internationalisation has concentrated mostly on established elite neighbourhoods like Kensington, Chelsea and Mayfair, but has tended to spread out more recently to other inner city areas with a close location to the centre. It has been reported that over 60% of new home sales in Central London currently stem from overseas investors. The major impact of this immense influx of capital has been a strong pressure on real estate prices at the top-end of the market. Thus, in 2011, a third of all money invested went into purchases of £5 million or more (Savills 2013, p. 8).

Over the last few years, the latter trend has become increasingly important for Barnsbury, too, as the opening of a Eurostar train service in neighbouring St Pancras station in 2007, with high-speed connections to Continental Europe, made the area very attractive to international buyers. The effect on the demand for housing in Barnsbury was more than considerable. In the words of a local real estate manager:

> Since the arrival of the Eurostar, the demand has changed completely. Previously we would have customers from the universities, the city [i.e. working in financial jobs, MB], from the media, and from legal services. Now we have even more people from the city, and more and more internationals. We have far more sales in top-end properties and the area is moving closer to the profile of Notting Hill or Kensington [which are well-established elite neighbourhoods in London, MB]. (Interview, 20 July 2014)

Penalty Renting

With this, we move to the topic of private renting, which is the second major change in the characteristic of gentrification in Barnsbury today (and, in fact across London). Thus, for London, while owner-occupation is still the dominant tenure, the share of private rentals, which has tended to fall since the 1960s and has remained at a level of around 14–17% for two decades, has grown enormously since 2001 and now has reached a level of 28% of all households. The Greater London Authority expects that this number will continue to rise.

This secular change is even more pronounced in Barnsbury. Here, as already discussed, private renting was the norm in 1961, including three quarters of all households. Throughout the 1960s and 1970s, this form of tenure was increasingly sidelined and replaced by a dominance of owner-occupation (mostly in the northern part of the ward) and council housing (in the southern part). Following the Conservative revolution in 1981, the picture changed again with council housing/social housing and private renting decreasing and owner-occupation rates rising until 2001. Since the 2000s, social renting has continued to decrease, but the upward movement of owner-occupation has stopped and the only tenure which is increasing is private renting, which by now has become the tenure for close to one-third of all households.

How is that to be explained? There are a number of both supply- and demand-side developments that can help to explain the trend.

On the demand side, the key drivers for the return of renting as a tenure discussed in the literature (see Wilson and Johnson 2013; Thomas 2014; Stephens and Whitehead 2013) are the following:

- *The declining availability of social housing*: As the supply of social housing has declined dramatically over the past three decades, more people are pushed into the private sector.

- *Rising house prices and more difficult access to a mortgage*: As house prices have increased, mortgage lending has become more difficult, resulting in fewer potential buyers getting into owner-occupation. As access to social housing has become very seriously restrained at the same time, this group is redirected into the private rental sector.

- *Increased flexibility in the labour market*: Employment has, in general, become more flexible and precarious, making it more difficult to access mortgages and sustain homeownership for a growing number of households across different industries and income groups.

- *Rising numbers of students, high rates of immigration and more singles*: The increasing number of students (which rose by 20% from 2000 to 2010) and the approximately 2 million immigrants that have entered the UK since 2000, both foster the demand for private rentals, as neither of these two groups usually has the track record or the deposit necessary for a mortgage and few qualify for social housing. In addition, later marriage and higher rates of separation have also fuelled demand for the private rented sector, as couples can more easily afford the high costs of ownership than single people can.

Taken together, these factors have led to a structural undersupply of both social and owner-occupied housing and have pushed people into private renting that would not have normally entered into this sector a decade ago. As a consequence, private renting has become the 'new normal'. It is reported that almost 50% of all households in the private rented sector are classified as 'high-income', and a third of these have children (Knight Frank 2014). Nevertheless, it would be a crude oversimplification to attribute the growth of private renting solely to the changing structure of demand. For a more comprehensive picture, a number of factors on the supply side need to be taken into account too:

- *The deregulation of rent control*: The creation of assured shorthold tenancies by the 1988 Housing Act has made renting out a flat or house easier than ever. As tenants hardly have any protection and tenancies can be terminated with a 2-month notice, landlords can easily increase rents and regain possession if they wish to. This has contributed to the confidence of landlords and catalysed their willingness to invest into this sector.

- *The introduction of buy-to-let mortgages*: In 1996, a number of mortgage lenders introduced buy-to-let mortgages to the UK market. This made mortgage much more accessible to landlords who previously had to rely on commercial loans with higher interest rates and low loan-to-value ratios.

- *A growing balance between the taxation of owner-occupation and private renting*: With the abolition of the mortgage interest tax relief in 2000, homeownership lost much of its traditional advantage with respect to taxation. The consequence has been 'a levelling of the playing field in favour of landlords which . . . may have discouraged some from entering homeownership' (Thomas 2014, p. 10).

Against this background, and in stark contrast to the 1970s, renting out has returned as a profitable business model. It has become so popular among homeowners that it has proven capable of fuelling a buy-to-let boom in the last decade. The outcome is a cottage industry of small-scale, part-time individual landlords investing for capital gains and managing their properties in their spare time (see Crook and Kemp 2011, Scanlon et al. 2014). For the growing number of private tenants, the return of the private rental sector, however, comes with quite a price tag.

First, as a result of the pressure on the market, private rents have increased tremendously over the last years. Median weekly private rent for Islington was around £433.25 in 2015 (London Datastore, https://data.london.gov.uk/dataset/average-private-rents-borough), compared with £124.59 in social housing and £111.85 in council housing. In Barnsbury, private rents start at £350 per week for a very basic two-bedroom property, but easily go up to more than £1100 for more convenient alternatives (see www.zoopla.co.uk, accessed 5 December 2016). Even with high incomes, this makes housing costs considerable. Private rents have increased greatly over the last years: between 2011 and 2015, median monthly private rents have increased by 15% across Inner London and by 25% across Islington (https://data.london.gov.uk/dataset/average-private-rents-borough, retrieved 3 March 2017, own calculations).

What is even more problematic is that under the current legislation in the UK, private tenants lack basic security over their place of accommodation. The rent that people pay is set by the landlord and is generally based on the landlord's view of what can be obtained in the current market situation.

A typical assured shorthold tenancy has an initial period of six or 12 months during which the renter can only be evicted for failing to comply with their tenancy agreement. After this, people can be evicted, when given two months' notice. The landlord does not need to provide any grounds for the eviction and there is no scope to legally challenge it. The legal situation of tenants is, thus, very insecure and in London it is not uncommon at all that tenancies are abolished after one year in order to make place for new lettings, together with rent increases. Current statistical data shows that 53% of private renters only stay up to one year in their home, compared with 17% in the owner-occupation sector and 13% in social housing (Housing in London 2014, table 6.1, https://www.london.gov.uk/sites/default/files/housing_in_london_2014_-_final.pdf, p. 104). In England, private renting is the most insecure tenure and often means constantly changing one's home and neighbourhood.

Due to the dominance of small, amateur landlords, the private renting sector is also notorious for rogue practices and inexperienced management. Disrepair, unlawful possession actions, harassment and problems with deposits are frequent in the sector, and their number seems to be rising in London (see London Assembly 2013; Copley 2014). As a consequence of their low security of tenure, tenants have severe problems in defending the minimum rights guaranteed to them by the law, as many landlords use 'retaliatory eviction' should tenants ask for repairs or cause other troubles.

To summarise, living in a private rented home in London is not usually a matter of choice, but a penalty for failing to achieve homeownership or winning in the social housing lottery. Private tenants tend to live in unsafe, unhealthy and poorly maintained properties; they pay an excessive amount of their income for rent; and they are fundamentally vulnerable to abuse by unscrupulous landlords and letting agents.

While renting has never been a pleasant experience in the UK, what is new is that not only poor families or transitory households are being pushed into this sector, but private renting has become more and more common among middle-class households too. Thus, while previous rounds of gentrification in the UK were based on a change from renting to owner-occupation, renting has now become part of the gentrification landscape.

Appendix B demonstrates this for the case of Barnsbury. The table lists the share of residents in the National Statistics Socio-Economic Classification (NS-SeC) Class 1 (Higher managerial, administrative and professional occupations) and Class 2 (Lower managerial, administrative and professional occupations) and the shares of different tenures in the respective 40 Output Areas of the Barnsbury ward, as well as in Barnsbury and London in general in the 2011 UK Census.

Comparing the individual Output Areas, an inverse relationship between the share of social housing and the proportion of better-off households is obvious. Geographically, this leads to a north–south divide of the ward, with lower incomes over-proportionately living in social housing stock in the south and richer households living in private rentals or in owner-occupation in the north. Nevertheless, even the most elite areas include minor shares of social housing, and at the same time, even the areas with the strongest concentration of social housing also include private rental and owner-occupied housing.

The overall share of private rented flats in Bransbury is above the London average. All Output Areas include private rentals. Thereby, the share is minimal in the Output Areas dominated by social housing, whereas it is highest[6] in areas dominated by the higher-class categories and owner-occupation. Even the most elite neighbourhoods with a share of more than 70% of residents in Class 1 and Class 2 now have a proportion of 17% to 35% private rentals. With this, traditional explanations for gentrification that identify it with a movement of middle-class owner-occupiers who risk their capital or invest their 'sweat equity' to fulfil their dream of owning a home and realising their lifestyle choices, are losing much of their former appeal.

There is a significant link between internationalisation of the population, the growth of higher managerial, administrative and professional occupations, and the upward movement of the private rental sector.[7] Of the 1168 newcomers to the ward between 2001 and 2011, 910 were born outside the UK, including 407 born in pre-accession European Union countries, 207 in North America and Australia, 124 in 'other Europe', and 101 in East and South East Asia – in other words, the growth of a high-price private rental sector seems to be driven by wealthy expats who are happy to rent, either for lifestyle reasons or because they know their stay will be temporary.

In summary, while gentrification was characterised by a transformation of rental flats into owner-occupied homes in the 1970s and a privatisation of municipal flats in the 1980s, now it includes a return of private renting even in the most expensive and most prestigious areas, going hand in hand with the internationalisation of demand.

In practice, the asset-maximising strategies sketched in this chapter and the growth of the private rented sector often work together. In business language, the main mechanism through which this is made possible is called 'leveraging'. The following example provides a picture of this.

A purchaser with enough cash buys a two-bedroom house or flat in Barnsbury at a price of £836 936 (which was the average asking price for a two-bedroom flat in Barnsbury in March 2017, according to www.zoopla.co.uk). They pay 20% of this as a deposit (£167 387) from their own pocket. The remaining £669 549 is financed with a mortgage. As interest rates are currently just a smidgeon above inflation, the liability resulting from this shouldn't be much of a problem. Calculating rather conservatively and assuming that the purchaser takes a mortgage at a rate of 3.5%, the monthly payment would be something like £488 per week. Rents for two-bedroom flats in Islington currently start at £325 for very simple flats (often in privatised council housing), but go up to more than £1200 a week for more luxurious apartments. Finding a tenant to pay £417 a week should, thus, not be difficult – which in effect means that the rental income that can be realistically expected would be sufficient to finance the liabilities resulting from taking a mortgage. If the purchaser manages to get a bargain and buy a flat at a below average price, get a better mortgage, or find a tenant who would pay a higher rent, they could even make a profit. The profit to be made from letting would, however, not be excessive and, in fact, profits made from letting only tend to be at a rate of 4–6% in London at the moment.

The good news, from the point of view of the investor, however, is that they can sell the property again. Metaphorically speaking, a buy-to-let investor is in the favourable position of having their a cake and eating it too. Calculating under the assumption that the rates of house price inflation stay at the current level (37% over a period of five years for Barnsbury, according to www.zoopla.co.uk) and assuming that the investor sells their property after 10 years, the price to be achieved for a two-bedroom flat bought for £715 000 in 2014 and sold 10 years later would be £1 244 100. The net gain achieved would, thus, be £529 100. If this related to the invested capital of £143 000, the profit made is 370%, or 37% per annum. The 'leverage effect' is, thus, considerable and makes real estate investment a much more lucrative investment than stocks in the current economic climate. Even with capital gains tax paid at the highest rate possible (28%), the after-tax return would be well above 25%. As there are numerous ways to shelter an investment against taxes (see http://monevator.com/avoiding-capital-gains-tax/), such losses are, however, rather unlikely.

From Value Gap to Super-gentrification

This changing political and economic environment has perfectly played together with the demand-side of gentrification. Thus, the closing of the value gap, the commodification of public housing and the upswing of asset-maximisation strategies, and buy-to-let investments did not bring gentrification to a point of saturation, but rather resulted in its modification and intensification. This development has been described as 'super-gentrification' by Butler and Lees (2006) and the changing composition of the ward population, indeed, provides

TABLE 4.2 Social class in Barnsbury based on Butler and Lees (2006) and UK Census 1992–2011 (own calculations)[8]

Social class	1981	1991	2001	2011
Higher managerial, administrative and professional occupations	6.9	12.0	32.0	44.0
Intermediate occupations	26.0	25.0	30.0	28.0
Skilled non-manual and manual workers	37.6	26.0	8.0	9.0
IV and V lower classes		37.0	30.0	19.0
Semi-skilled/ unskilled	25.8	11.0	13.0	
Unemployed, inactive and other	3.7	26.0	17.0	–

evidence of an intensified 'elitification' of the population. More and more, even middle-class households that dominated earlier rounds of gentrification are being displaced by richer residents (and the lower classes are being ousted and relegated to the remnants of social housing; see Table 4.2).

Keeping in mind the difficulties in comparing data from different census surveys, the UK Census clearly displays a massive growth of higher managerial, administrative and professional occupations, which have become the dominant class in Barnsbury. Over the years, the 'managerialisation' of the neighbourhood has increased, and Barnsbury's population has become continuously richer, more international and with ever more people working in the financial sector. At the same time, workers have next to disappeared from the area, while the share of unemployed and economically inactive residents (who live in the remaining social housing stock) remains considerable.

If the gentrification of Barnsbury is studied over the last decades, a number of differences are palpable, which clearly distinguish its current appearance from the past:

1. Gentrification has become more elite and more globalised. Larger parts of the autochthon middle-classes are priced out, and especially potential first-time buyers are excluded from living in a neighbourhood like Barnsbury. In super-gentrifying neighbourhoods like Barnsbury, exclusionary displacement (Marcuse 1986) has become the main mechanism of gentrification, at least in the owner-occupied sector, and even the social groups that would traditionally be regarded as gentrifiers are increasingly exposed to it.

2. The more that middle-income households can't afford to buy a property, the more important private renting becomes. If gentrification in the 1960s and 1970s was driven by a change from renting as the dominant

form of tenure to owner-occupation, the last 10 years have seen the opposite movement. With this, more and more gentrifiers are subjected to the very unfavourable conditions typical for the private rental sector in London. In the past, this experience has been mostly reserved for the lower classes.[9]

3. The capital gains that can be made from withdrawing equity from housing owned have set in place powerful spill-over dynamics. More and more, asset-rich middle-class households cash in, sell their homes to the super-rich and move to a more affordable neighbourhood, driving prices there upwards. This is partly enabled by the extraordinary situation that housing in London is highly attractive as an investment for foreign buyers, which has accelerated demand in the market. The combination of continuous new supply through equity withdrawal, working together with inflated demand, however, is a cascading effect by which gentrification pressure is directed outwards and a chain of displacement effects is set in place.

The spiralling up of gentrification has been accompanied by a series of mutations in which the political economy underlying the provision of gentrifiable housing has more than once changed fundamentally. Thus, in its early days, gentrification was based on capturing the value gap between renting and owner-occupation that was made possible through a very specific combination of inflation, British tax laws and rent regulations. The drivers of gentrification at that time were speculators in the 'flat break-up market' paving the way for middle-class owner-occupiers who purchased homes for personal use.

In a second phase, this orientation on owner-occupation was continued, but by exploiting the state-induced rent gap, which was opened up through the Right to Buy policy. The privatisation of council flats imposed the potential ground rent established through the closure of the value gap proceeding since the 1970s on formerly public housing, thus, making it possible to create huge profits from previously outside-the-market stocks. This was all the more dramatic, as it coincided with the 'Big Bang', i.e. the opening of the London Stock Exchange to international investors, which injected a substantial volume of purchasing power. Thus, while demand grew, so did supply – not only by expanding the total stock, but by opening up segments of it that had been fenced off from the market to be taken over by potential gentrifiers.

In a third phase, subsequent price inflation, the deregulation of rents and the introduction of new financing instruments worked together to make gentrification a strategy through which property owners could 'escalate' their assets, making them a 'wealth machine' (Edwards 2016, p. 234) both for those lucky enough to having gambled in this process early on and, increasingly, for global investors. The effect was an ongoing 'eliticisation' of gentrification, accompanied by a return of the private rental sector, making even super-gentrifiers more vulnerable to speculation and asset-maximising strategies.

What does this story tell us conceptually? The easy answer would point to the rent theory and claim that in all these different environments, potential ground rents were higher than what was actually realised, so capital was invested to profit from this difference. As a consequence, prices went up and income-poor residents were driven out. While this is certainly the case, the story to be told is much richer and more complex. Analysing the progress of gentrification in Barnsbury over time, it becomes clear that gentrification is not only an outcome of a difference between actual and potential ground rent, but both the emergence of this gap, its geography, the way it operated and the opportunities to profit from it have fundamentally altered throughout the last five decades. Sticking with the metaphor of gaps, numerous gaps can be identified that worked in different conditions, sometimes in parallel, sometimes sequentially.

Thus, a first value gap arose when tax policies, mortgage regulations and rent laws interacted in a way that made the transformation of private rentals to owner-occupation profitable. This classical value gap, which has been intensively studied by Hamnett and Randolph, resulted in the displacement of low-income tenants and the takeover of the vacated stock by in moving owner-occupiers.

With the right to buy, a second gap emerged, which could be called a privatisation gap. Similar to the value gap, it rested on the change of tenure, yet this time from council housing to owner-occupation. The crucial characteristic of this gap is that it was based on the deliberate commodification of a decommodified stock by the national state.

A third gap, working throughout the period examined but gaining more and more relevance since the turn of the millennium, can be termed the inflationary gap. It rested on the financialisation and globalisation of property markets. In this context, yields from land titles are increasingly traded as a financial instrument and, at least to some degree, potential ground rents can be portrayed as 'derivative rents' (Haila 2016, p. ixx). In summary, this development has pushed soaring price increases in London. Obviously, this trend works in favour of (outright) homeowners and results in a continuous rise of the potential ground rent – even without further investment. The consequence of this gap is both 'spatially displaced demand' (i.e. a ripple effect on gentrification) and super-gentrification excluding large parts of the middle-class.

A fourth gap could be called a 'rental deregulation gap'. This has developed only in the last two decades. It is based on the deregulation of private rental contracts and the introduction of buy-to-let mortgages and has made private renting or letting a lucrative business again. As shown in this chapter, the operation of this gap interacts closely with the inflation of house prices. Its effect is the return of private renting, which has simultaneously made the classical value gap disappear and pushed gentrification into the private rental market.

The regulatory environment of gentrification has, thus, changed considerably over time and has defined diverse contexts around which capital revalorisation was able to take place.

Notes

1. There are no precise boundaries for what can be thought of as Barnsbury. For reasons of practicality and data access, this study focuses on the Barnsbury Ward, i.e. on the administrative unit that roughly includes the area between Offord Road and the trainline in the north, Pentonville Road in the south, Upper Street and Liverpool Road in the east, and Caledonian and Hemingford roads in the west, thus, leaving out a number of blocks to the northeast and the northwest that could also be considered a part of Barnsbury in a physical sense, but including social housing estates south of Copenhagen Street.

2. While 'value gaps', as they were described by Hamnett and Randolph, and 'rent gaps', as described by Smith (1979), seem to be fairly similar concepts at first glance, there are a number of differences between the two. In short, the value gap explores the opportunities of achieving higher revenues through tenure change, whereas the rent gap is a more general theoretical category that operates at the level of the ground rent. Thus, the value gap addresses a rather specific regulatory and institutional context, whereas the rent gap is more general and at the same time place-specific (see Clark 1992). As exploiting the opportunities of a tenure shift unavoidably affects the ground rent on a piece of land, a closure of the value gap also 'entails a partial closure of the rent gap' (Clark 1992, p. 20). Both theories should, therefore, be seen as meshed rather than in opposition. Often, the value gap has been essentialised in the literature. Loretta Lees (1994, p. 207), thus, sees the value gap 'as a more appropriate theorisation of processes of gentrification in England', resulting from different landholding systems. Van Weesep and Musterd (1991, p. 15), in contrast, have linked the value gap to the impact of urban planning in European cities, which would make fast increases of housing prices difficult, and Dangschat (1989, cited in Clark 1992, p. 24) has even posited that the rent gap would be a more appropriate theorisation for the 'zone in transition', hence, more applicable to the USA, whereas the value gap fits better in the former bourgeois areas of European inner cities.

3. See Chapter 3, note 4.

4. As Paccoud (2016) points out, this is in contrast to the traditional bias of much gentrification research on 'owner gentrifiers' and their liberal values of producing for personal use, instead of profit.

5. The MSOA (Middle Super Output Area) Islington 2017 only partly overlaps with the Barnsbury ward. Following changes in the census district definition, the Barnsbury ward is now divided into four different MSOAs that partly include sections of other wards too. For purposes of illustration, I have selected the Islington 017 MSOA, which includes the most prestigious parts of Barnsbury.

6. E0013422 and E00174843 must be counted as an exception here. One of these is dominated by a student housing complex that provides student apartments in the form of rentals and the other is a shopping street with residential use in the upper floors.

7. I owe this insight to Antoine Paccoud from the LSE who kindly pointed me towards this.

8. Note: Between 1991 and 2011, changes in the division data were made, making exact comparison between the years 1991, 2001 and 2011 impossible, and skewing the percentages. Thus, the subdivision of the classes varies across different samples. For example, the class 'skilled manual workers' in 1991 is divided into non-manual and manual workers, while apparently they are both considered to be skilled manual workers. A second factor that complicates a comparison is the difference in measurement between 1991 and 2011. For the years 1981 and 1991, the data for social classes was displayed by household. In 2001, the data was displayed for both the exact number of persons and for the economically active heads of household. For the year 2011, the only available data concerned the exact number of persons. In addition to this, the boundaries of the Barnsbury Output Area were subject to changes. In summary, the data needs to be handled with care.

9. Traditional elite areas like Knightsbridge or Mayfair are to some degree an exception to this rule. They have for a long time been inhabited by the British aristocracy who generally regarded their country house as their main home, but typically spent several months of the year in the capital.

CHAPTER 5

Prenzlauer Berg: Gentrification Between Regulation and Deregulation

Prenzlauer Berg[1] is an old inner city district in East Berlin that lies in the immediate vicinity of the city centre. Being built up at the turn of the nineteenth century, the area experienced considerable neglect under East German state socialism. In the 25 years following the fall of the wall, the neighbourhood has seen a close to total renewal and massive population exchange. Within Germany, Prenzlauer Berg is widely known and is usually seen as a showcase example of gentrification. Journalists have even used the term *Prenzlauerbergisierung* (Prenzlauerbergisation) when trying to describe urban changes elsewhere. The changes in the neighbourhood have been widely studied and documented both in academic writing (e.g. Bernt 1998; Häußermann and Kapphan 2002; Häußermann et al. 2002; Bernt and Holm 2005, 2009; Dörfler 2010; Bernt 2012; Bezirksamt Pankow 2015) and in fiction, and the theme is so omnipresent that it even features in a weekly cartoon in the regional newspaper.

In stark contrast to public images, the transformation of Prenzlauer Berg has taken place in a widely state-regulated environment. The vast majority of flats in the area are rentals and, as such, they are subject to the complex system of regulations discussed in Chapter 3, in the section The German Experience. In addition, most of the neighbourhood had the status of an Urban Renewal Area for more than 20 years. This set in place numerous local regulations limiting the leeway for renovation schemes and rent increases. Moreover, massive direct and indirect subsidies were channelled into Prenzlauer Berg: €670.1 million (see Bezirksamt Pankow 2015, p. 317, own calculations) have been spent by the city of Berlin to directly support the renovation of buildings and infrastructure with approximately an additional €1 billion of tax relief for investors. Being an Urban Renewal Area also meant installing special administrative units and commissioned agents.

The Commodification Gap: Gentrification and Public Policy in London, Berlin and St Petersburg, First Edition. Matthias Bernt.
© 2022 John Wiley & Sons Ltd. Published 2022 by John Wiley & Sons Ltd.

The regeneration was, therefore, supervised by 30–70 (numbers have changed over time) administrative officers, state commissioned planners and tenant advisers (Bernt 2003, p. 241). In this environment, urban renewal was clearly a state-subsidised, state-regulated and state-guided undertaking. The declared goal of these public efforts was simultaneously to support the renovation of neglected building structures and to make sure that this renovation would not lead to a displacement, protecting the social mix in place. Thus, while renewal proceeded in a widely state-regulated environment, there is a massive discrepancy between the declared goals of this massive public intervention and the outcomes achieved. How could this be the case? As I will show in the this chapter, the conditions for, and the modes of, public intervention have always been contradictory. They have also changed over time. In this sense, five phases can be distinguished.[2]

From Plan to Market

When the Berlin Wall came down in 1989, state socialism had left most of Prenzlauer Berg in neglect. Around one quarter of the 45 000 apartments in the later formally declared Urban Renewal Areas were seriously damaged, and most of the rest were still in bad condition; nearly all residential buildings lacked central heating and half of the housing stock did not have a separate bathroom. The need for renovations was so pressing that the neighbourhood acquired the reputation of being Europe's largest Urban Renewal Area at that time. Nevertheless, as an outcome of socialist housing policies, and in stark contrast to the Barnsbury case, the social structure of the population was fairly mixed, with professors and sanitation workers living next door to one another. What made Prenzlauer Berg peculiar, however, are not so much the dimensions of abandonment and neglect, but the political prerequisites for the urban renewal yet to come.

First, and different from many of its fellow West German states, West Berlin had developed a model of a socially balanced 'Careful Urban Renewal' (*Behutsame Stadterneuerung*), which became the official urban policy of its municipal government during the 1980s. This model responded to powerful local tenant and squatter movements and aimed to reconcile physical renewal with a socially balanced development, promising, among other goals, to preserve the existing social structures against displacement due to rent increases (for more details on the history of Berlin's urban renewal politics, see Bodenschatz 1987; Bernt 2003). Based on a blend of Keynesian welfare politics, mixed with a post-modern appreciation of culture and authenticity, this Careful Urban Renewal model was almost entirely financed by public subsidies. In fact, in the Urban Renewal Areas of the 1980s, close to 100% of the refurbishment was publicly funded (see Bernt 2003, p. 67), and an important side effect of this support were extensive rental and occupational obligations that had to be accepted by the landlords receiving the subsidies. As a consequence of this massive public support, the development of rent prices in Urban Renewal Areas was bound to very affordable levels for up to 30 years.[3]

This West Berlin heritage played an immense role in defining the agenda for urban renewal in East Berlin in the period following reunification. At that time, there was a wide political consensus in Berlin that the local government had a responsibility to guide urban renewal and protect the residents from displacement and individual hardships. Early attempts from within the state bureaucracy aiming at cost-reduction and budgetary discipline were immediately stopped by a coalition of social movements, tenant organisations, planners and left-wing politicians who made sure that the strategies for urban renewal in Prenzlauer Berg would, by and large, follow the script followed in West Berlin. Thus, when the first Urban Renewal Areas were declared in East Berlin in 1993, their formal declaration came together with a new document, Guiding Principles for Urban Renewal in Berlin (*Leitsätze zur Stadterneuerung in Berlin*, Senatsverwaltung für Bau- und Wohnungswesen 1993), which included the following ambitious goals:

> 3. The renovations must be focused on the needs of the inhabitants. The renewal measures will be organised in a socially responsible way. . . . In the case of privately financed renovations, it is imperative to avoid negative consequences that would collide with the social intentions of urban renewal, as well as the preservation of the social structure. . . . it is mandatory to avoid: the displacement of low-income groups, the acceleration of processes of residential segregation . . . as well as individual hardships for adaptable households. In principle, urban renewal should allow the inhabitants to stay in the area. The rent increases shall, therefore, be focused on the capacities of the inhabitants.
>
> (own translation)

In stark contrast to the role of the state described in much of the gentrification literature, the state of Berlin, thus, made it a public goal to avoid displacement and protect low-income households. While place competition and entrepreneurialism increasingly dominated the overall urban development agenda (see Helms 1992; Krätke 1995), urban renewal maintained the welfarist orientation inherited from the 1980s.

Yet, the financial basis on which this was to be achieved became more and more difficult. In Kreuzberg, the focus of renewal activities in West Berlin in the 1980s, Urban Renewal Areas had included around 10 000 flats – in Prenzlauer Berg alone the number was close to 33 000 (see Bernt 2003, p. 67; Bezirksamt Pankow 2015, p. 308; own calculations). Across all East Berlin, 315 000 flats were to be renovated (Senatsverwaltung für Bau- und Wohnungswesen 1993). The requirements in terms of finances were, thus, incomparably higher than in the 1980s, and it was clear from the start that urban renewal could hardly proceed as an almost completely publicly financed undertaking (as happened in West Berlin). This proved particularly true as the financial situation of Berlin became more difficult after reunification. Regarded as the showcase of the West, West Berlin had always been massively supported by the federal government with special financial allocations during the Cold War which had allowed for many social, cultural and ecological experiments, but also for a biotope of intensive real estate speculation based on claiming public subsidies. It has often been said that the Berlin Wall had been something of a life insurance for West Berlin's economy, as it justified extensive public transfers from West Germany, and this was especially true for the field of housing. After reunification, this insurance was abandoned and the special allocations for West Berlin were successively reduced and eventually cut. The outcome was a rampantly growing deficit in Berlin's budget that increased from 1.6 billion DM in 1991 to 10.7 billion in 1995 (Vesper 1997, 1999). Thus, while Berlin aimed to foster Careful Urban Renewal at unheard-of dimensions, making sure it would follow social goals, the resources for this shrank rapidly.

The consequence was a successive reorientation to private investors, who were to fill the gaps left by a lack of public resources. In this sense, the Guiding Principles declared:

> Given the restrictions for public support and the extreme growth of public duties in the eastern boroughs succeeding with urban renewal will depend on mobilising private investment. . . . 9. The necessary renewal of old buildings can only be achieved when investments from the side of the owners are activated and the measures are increasingly financed through private capital.
> (Senatsverwaltung für Bau- und Wohnungswesen 1993)

Without making this very explicit, these subclauses included a wholesale reorientation. While urban renewal in the past had relied upon public and quasi-public companies buying properties and implementing renewal in the way aligned with predetermined, publicly declared goals, the new guidelines have given up on this tradition and aimed at achieving the goals of urban renewal in collaboration with the existing owners.

In retrospect, this strategy turned out to be built on sand, as ownership structures in East Berlin changed rapidly after reunification and did so in a way that was not beneficial to the claimed goals at all. To a large part, this was caused by the encounter of socialist housing policies in the former German Democratic Republic (GDR) with the specific German pathway towards transition after reunification. When the wall fell, in most of the GDR private homeownership was close to irrelevant, as a consequence of the socialist planned economy. The lion's share of the housing stock was managed by state administrations and only a minor part was still owned by private persons. No matter if state or privately managed, rents, investment opportunities and occupation were strongly regulated and even the inhabitants often couldn't tell whether they lived in a state-owned or a private flat. Privately owned property – which the Senate was increasingly to rely upon – thus, had yet to be established in the East of Berlin.

The way this was done was by the restituting properties to their original owners. East Germany had seen two waves of expropriation of property owners – first under the Nazis and then by the communist government – and, moreover, many landlords had simply abandoned their property in the GDR as they saw no use in it. The Unification Treaty between the GDR and the Federal Republic of Germany ruled that all these properties had to be given back to the original owners, or their heirs.[4] This restitution had two important consequences. First, property rights were re-instituted as the main determinant of urban development. In stark contrast to most other post-socialist countries, sitting residents were practically excluded from participating in this privatisation. Secondly, once the original owners were given their property back, between 70% and 90% of them immediately sold it on to professional housing companies and developers (Dieser 1996, Reimann 2000). The outcome was a professionalisation and commercialisation of ownership structures. As Häußermann (1995, p. 13) put it then:

> It is not the old dynasties of craftsmen and traders who become the owners of inner-cities again, but Real Estate Investment Funds, international realtors and speculators of all kinds. Real Estate capital . . . gains a role in urban development which has not yet been known in German cities.

With high costs for renovation (resulting from serious neglect of the building stock under socialism), average house prices of 800 DM per square metre and interest rates of around 7%, this led to an immense pressure on rents, and it soon became clear that rents would need to rise far above the existing level to make an investment lucrative. Restitution, thus, fuelled a speculative bubble that exerted a high pressure on existing rental and sales prices. Thus, while the guidelines aimed to implement Careful Urban

Renewal in collaboration with private investors, private investors were interested in achieving quick profits and had acquired their properties at prices that exerted a strong pressure on rent increases. This mix was bound to go wrong and, as the head of a renewal company expressed it in an interview, 'this basic structure made the renewal strategy adopted increasingly obsolete'.

In the first years following reunification, this was not yet very obvious though. Resulting from the still considerable weight of inherited political traditions, as well as the gap between the formal onset and actual realisation of restitution, renewal remained a major publicly subsidised endeavour. More than 60% of all renewal work between 1992 and 1995 was financed by and organised in public programmes that tied the properties concerned to rental and occupational obligations. In summary, this affected around one-sixth of the entire housing stock of Prenzlauer Berg.

Rolling out the Market, Weakening Public Control

In the second half of the 1990s, Berlin's urban renewal policy changed successively. With the unfolding of the municipal budget crisis, the share of publicly financed refurbishments dropped to about one-third of all renewal activities between 1996 and 1999, and a first phase of privately financed refurbishment gained ground (see Table 5.1).

Together with the upswing of private investments came the first cases of ferocious rent increases and displacement. Motivating private investment, while at the same time avoiding displacement and protecting low-income

TABLE 5.1 Renovated flats in Urban Renewal Areas in Prenzlauer Berg 1994–2001 (Holm 2006, p. 185)

	No. of renovated flats	No. of reno-vated flats that were subsidised	% of renovated flats that were subsidised
1994	811	811	100%
1995/1996	5 907	3 404	58%
1997	1 468	787	54%
1998	2 300	1 925	48%
1999	2 356		
2000/2001	4 087	1 609	39%
Total 1994–2001	16 938	8 536	50%

tenants became more and more contradictory. In this situation, tenant organisations and neighbourhood groups demanded more protective policies be set up and an intensive search for new strategies began. After a period of back and forth, this resulted in the introduction of local rent caps for privately financed renewal activities in 1995. These were designed in a way to cap rents after renovation and modernisation for a certain period (depending on the borough, for between three and seven years) at a socially acceptable level of around 6–9 DM per square metre. Special, mandatory permissions for all refurbishments in designated Urban Renewal Areas formed the basis for these rent caps.

In reality, however, most owners found ways to bypass these regulations and realised considerable rent increases. For example, they easily identified the weakest point in the rent regulation (that, in practice, it only applied to existing tenants, who already lived in their flats before the refurbishment started), and paid up to 20 000 DM to compensate tenants for moving out. By 2002, the proportion of existing tenants for whom the new rental prices were capped, however, had decreased to around a quarter (see Table 5.2).

The growing rate of new tenants was equalled by a rising rent level. Thus, in modernised housing, half of the tenants who moved in after renovations paid almost twice as much as previous tenants with rental caps in place (Häußermann et al. 2002). Thus, although the introduction of rent caps for private refurbishments constituted an innovation implemented at the local level, which was very much in the interest of low-income tenants, in the long run, it proved to be a step towards weaker forms of regulation.

Examining the effect of private renovations on rent levels and the social composition of tenants, it is important to keep in mind that the majority of private refurbishment activities did not rely on their own resources during this time. Quite the contrary, instead of using direct municipal subsidies linked to rental and occupational obligations, from the mid-1990s onwards, the majority of refurbishment projects relied on special depreciation possibilities guaranteed in the federal Development Zone Act (*Fördergebietsgesetz*). Until 1996, this form of depreciation allowed up to 50% of refurbishment

TABLE 5.2 Percentage of sitting and new tenancies after renovation in Urban Renewal Areas in Prenzlauer Berg 1995–2002 (Holm 2006, p. 310)

	1995	1998	2000	2002
Sitting tenancies	60	50	35	25
New tenancies	40	50	65	75

costs in the first year of investment to be offset against one's taxes, with this proportion reduced to 40% until 1998/1999. These high, indirect subsidies made the refurbishing of old housing stock extremely lucrative for investors with a large taxable income, especially if costs were high and rents low, since the costs of investment could be transformed into tax savings for the investing partners involved. As the balance sheet of the investment could be evened out by tax advantages, investors could afford to forego high rental income for a while, as well as to build in areas without apparent affluent demand.

This had a considerable effect on the geography of renewal activities, as investments took place largely irrespective of location and covered a wide territory. Andrej Holm and I described the consequences for gentrification at this time (Bernt and Holm 2005, pp. 119–120 and 112) as follows:

> . . . direct and indirect subsidies resulted in investments being widespread rather than concentrated and the rent increase (at least for existing tenants) is slowed down by a series of regulations. Urban change in Prenzlauer Berg is therefore puzzlingly split. On one hand, in the case of those dwellings where refurbishment is carried out with private money and where the rents for new tenants are freely negotiated, gentrification and the displacement of poorer households shows classical features . . . On the other hand, a supply of substandard housing has remained throughout the district for a long period of time which is still being used by lower-income groups. As a result, poorer and wealthier sections of the population are living side by side, delaying the transition from a pioneer phase of gentrification. . . . Instead of having a clear 'frontier', gentrification in Prenzlauer Berg has had a restricted scattering of investment across the neighbourhood.

Since 2000: Privately Financed Refurbishments, Condominium Boom and No Regulation

After the turn of the millennium, the course of urban renewal in Prenzlauer Berg changed again. The opportunities for tax deductions ended in 1999 and, as a consequence, the previous business model that renewal had been based upon became outlived. In addition to less favourable opportunities for tax deductions, the economic and institutional environment changed considerably with regard to at least three other issues, and this resulted in a modification of renewal strategies.

First, following a scandal around Berlin's state-owned bank, the city slipped into a dramatic financial crisis. The gambling away of several billion

Euros used as security for private real estate transactions through a publicly owned bank led to an extreme budgetary emergency (*Haushaltsnotlage*) and a new Red-Red governing coalition came into power and introduced severe austerity policies. The outcomes were drastic budget cuts and wholesale retreat from numerous tasks and funding programmes, as well as the privatisation of public housing stock and communal infrastructure companies. One of the first victims of this new policy was publicly financed urban renewal, and existing funding programmes in this sector were completely abolished in 2001. By most observers, this turn was regarded as a decisive blow, and yet another nail in the coffin of an already stressed Careful Urban Renewal. If in the early 1990s the strategy was largely based on public subsidies and the late 1990s had already experienced a successive weakening of this approach, in 2001, it was given up completely.

As a consequence, more and more refurbishment activities based their calculations on the transformation of rental housing into single ownership, bringing to mind the 'flat break-up market' described for London in the 1970s by Hamnett and Randolph (Hamnett and Randolph 1984, 1988). In this strategy, investors developed construction plans for single-ownership apartments and sold not yet existing (i.e. de facto 'fictitious') condominiums before refurbishment actually started. This strategy made refurbishments financially attractive again and allowed a sufficient return on investment, yet at a cost to existing tenants. The major problem lies in the fact that in order to market the planned condominiums successfully, the tenancy needs to change and the possibilities for existing tenants to remain have to be reduced to a minimum. Within this framework, the displacement of existing tenants reached peaks that were not previously thought possible, and the existing tenants' rate decreased to less than 25%. This model of renewal proved to be very popular among investors between 2000 and 2010, so that up to one-third of all housing units in the renewal areas of Prenzlauer Berg were transformed into condominiums (see Table 5.3).

This comparably high share of condominiums has considerable implications for the population mix. Thus, while it is impossible to determine from official statistics whether a freehold flat is occupied by the owner or let, it is commonly estimated that the majority are rented out again. This assumption is also supported by a study conducted on the Urban Renewal Area of Helmholtzplatz (ASUM 2012, pp. 41f.), which reports that only around a quarter of all converted flats are occupied by their owners, whereas the rest are rented out. Thereby, tenants in individually owned apartments pay higher rents and they have higher disposable incomes and higher qualifications than the average population in the neighbourhood. The problem here is that, as discussed in Chapter 3, in the section The German Experience, converting

TABLE 5.3 Number and percentage of individually owned apartments in Urban Renewal Areas in Prenzlauer Berg (Bezirksamt Pankow 2015, p. 311)

	No. of flats in total	No. of flats that were condominiums	% of flats that were condominiums
Kollwitzplatz 2009	7 072	1 424	20.1
Winsstrasse 2011	5 200	1 903	36.6
Teutoburger Platz 2013	4 855	1 598	32.9
Bötzowstrasse 2011	3 559	1 109	31.2
Helmholtzplatz 2015	13 480	3 297	24.5

rental units into individually owned flats undermines the security of tenure, as landlords can easily claim that they need the property for personal use. This enables the evacuation of apartments and the achieving of new tenancies at higher rates. In other words, with this approach, it is possible to bypass rent regulations, which mainly work for existing contracts, and achieve market rates. The conversion of rental homes into privately owned homes, therefore, does not necessarily equal a change of tenure – but next to always results in rising rents and new residents, thus indicating some sort of buy-to-let gentrification.

It should be emphasised that (while of lesser significance than in the past), this operation has also been backed by tax subsidies offered by the German state, both on the supply and on the demand-side. As §7h of the German Income Tax Act enables investors to write-off up to 9% of the building costs annually in the first eight years after construction has ended, followed by 7% for another four years in formally declared Urban Renewal Areas. Investment in such areas was comparably lucrative, especially when taking into account globally low interest rates as we have seen since the turn of the millennium. Between 1999 and 2015, this made up a sum of €594.3 million (Bezirksamt Pankow 2015, p. 318). Moreover, purchasing an apartment is also subsidised for the buyer through special allowances (*Wohnungsbauprämie, Eigenheimzulage*), low-interest credits (*KfW*) and opportunities for tax deductions. As in the previous phase, gentrification was, thus, subsidised with public resources – yet not from the budget of Berlin, but at the cost of the general taxpayer, and with no obligations whatsoever on the receiver.

Although the consequences of condominium conversion were detrimental to the still official policies of Careful Urban Renewal, they were not fought against, but largely supported by Berlin's Social Democrats, which led the Urban Development Department of the city at that time. The background for this was the belief that higher ownership rates could assist in stabilising stressed neighbourhoods and provide incentives for middle-class families to move into poorer inner city areas. On this basis, the Guiding Principles for Urban Renewal in Berlin were changed in 2005, and supporting the in-migration of 'stabilising segments of the population' (Senatsverwaltung für Stadtentwicklung von Berlin 2005) into Urban Renewal Areas was made an official policy. While Berlin's budget was too stressed to actively support these efforts, the owner-occupation strategy was at least not fought against, and whereas other German cities had implemented a ban on transforming rental flats into condominiums since 1998, this opportunity was not taken by Berlin's government.

The development proved to be particularly problematic as the rent caps introduced for privately financed renovation activities in 1995 were eventually declared illegal by the German Federal Administrative Court in 2006. With this, the court upheld a claim of a property owner in Berlin-Friedrichshain who had fought the rent limitations demanded by the borough for years. It argued that local rent caps would not be in line with the general legislation on rents and claimed that individual municipalities' attempts to bypass this system on the basis of planning legislation was illegitimate.

This delivered the final blow to the inherited urban renewal strategies. In the mid-2000s, protecting low-income segments of the population against rent increases and maintaining mixed income structures were still an official goal of renewal – yet they could not be bought in exchange for subsidies or determined by regulative instruments, nor could landlords be convinced to follow these strategies voluntarily. Increasingly, renovations resulted in expensive apartments, most of which were for sale and could rarely be afforded by the average resident in the area. Whereas gentrification pressure became stronger than ever before, local regulations aimed at the protection of low-income tenants had become close to non-existent.

New Build Gentrification and Energy Efficient Displacement

On this basis, in the mid-2000s, the picture of gentrification in Berlin was as described here. Most apartments had been renovated. Of these, between 10% and 20% of the renovations had been supported by public subsidies, resulting in comparably affordable rents. The rest were more expensive – and becoming less affordable year by year (see ASUM 2012, p. 29). Even houses that were renovated with private means, however, often contained low-income residents who had profited from rent caps in the 1990s and, thus, enjoyed a comparably low base upon which for further rent increases would be determined.

In this situation, the previously existing rent gap had nearly ceased to exist and there remained only two ways in which capitalised ground rents could still be pushed up.

First, while most of the houses were renovated, there still existed numerous undeveloped lots and vacant industrial sites in the area. Starting in the mid-2000s, these successively became the place for new building projects in the upper market segment of the market. Naturally, the geography of these projects followed the location of lots that could be developed – resulting from this, the share of new buildings differed considerably between different Urban Renewal Areas (see Table 5.4). In any case, the share of new built apartments proved to be significant.

Often, these new buildings were erected in larger complexes of up to 100 flats or more. The 780 units built at the Urban Renewal Area Teutoburger Platz, thus, include 129 units at the luxury estate of Marthas Hof; the 370 units at Bötzowviertel entail more than 200 flats at the Schweizer Gärten project; the 870 units at Kollwitzplatz include 98 units of the Palais Kolle Belle project; etc. While there is no comprehensive survey of the social composition of the residents of these new flats, a study by Andrej Holm (2010) provides a rather clear picture about the kind of gentrification implied. Nearly all flats built in these developments were sold as condominiums, for an average price of €3700 per square metre which is more than double the average price in Berlin. In addition, most of the apartments are extremely spacious and have net dwelling areas of 280–450 square metres. Purchase prices of

TABLE 5.4 New built housing units in the Urban Renewal Areas of Prenzlauer Berg (Bezirksamt Pankow 2015, p. 308; ASUM 2012, p. 23)

	No. of housing units (2012–2015)	No. of newly built housing units	% of newly built housing units
Kollwitzplatz	7 072	870	12
Winsstrasse	5 200	640	12
Bötzowstrasse	3 559	370	10
Teutoburger Platz	4 855	780	16
Helmholtzplatz	13 880	600	4

more than €1 million were the rule rather than the exception. Consequently, the population structure was very elite, and included a high percentage of international professionals, managers and well-known media-stars. On this basis, Holm equated the boom of luxury developments in the 2000s with super-gentrification, as it has been reported in New York City and London.

Again, this development has been facilitated by public policies. Most of the large-scale projects were built on land previously owned by the state or state-owned companies and, therefore, could not have happened without privatisation. As in London's privatised council stock, the rent gap closed here is state-induced. In addition to land supply, the boom of newly built developments was also supported by a deregulation of building regulations in 2006, which enabled far higher densities and lower shares of green space.

The second development pushing forward gentrification in the already fairly gentrified Prenzlauer Berg neighbourhood, is the renovation of already refurbished buildings, aiming at higher standard equipment like a second bathroom, a second balcony, modified floor plans or floor heating. Often, these projects go hand in hand with a conversion to individual ownership. From the perspective of the investor, these improvements make sense because they facilitate using the potential included in the – by now – prestigious location, and achieve exclusive rents for exclusive standards. What was new then was that the supply did not target middle-class households, but even these were increasingly put under pressure. Again, it took years until the borough found ways to at least modify this trend. For this, the borough massively expanded Milieux Protection Areas (*Milieuschutzgebiete*), either placed upon already existing Urban Renewal Areas, or as new areas or as an expansion of already existing areas. This enabled them to establish construction activities subject to prior approval, in order to determine a catalogue of special features that would be capable of considerably increasing the comparative rental value and rent increases, like marble ceilings, second balconies, etc. As Figure 5.1 shows, this tactic is certainly not limited Prenzlauer Berg but has become a fairly common instrument applied in many boroughs in Berlin.

As the declaration of Milieux Protection Areas needs careful justification and approval by upper administrative levels and includes complex legal issues, this process took a couple of years. Moreover, there are two fundamental difficulties that limit the effectivity of this instrument. First, the borough is not entitled to outlaw what is called 'contemporary equipment features' (*zeitgemäßer Standard*) in German legal materials. While it is, thus, possible to ban golden water pipes, stopping the installation of an escalator or a modern heating system is all but impossible. The problem here is what I call the 'modernisation gap' (see Chapter 3, in the section The German Experience), i.e. that the opportunity enshrined in German rent legislation to pass the costs

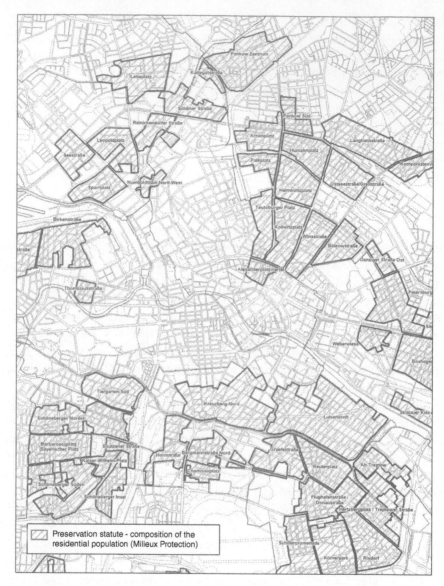

FIGURE 5.1 Milieux Protection Areas in Berlin (2016). Altogether, close to 900 000 residents live in Milieux Protection Areas today (Geoportal Berlin / Erhaltungsverordnungsgebiete §172 BauGB).

for new features to the tenants and increase the rent by 11% of the costs per year. Even with average standard increases, this regularly justifies rents that are way beyond affordable for lower-income groups. Preventing luxury standards through Milieux Protection is, thus, capable of capping rent increases at the very top end of the market and limiting the turnover from gentrification to super-gentrification, but it hardly provides much protection for the groups of residents that are the most vulnerable to even more moderate rent increases. Second, the introduction of new legislation on energy efficiency in the housing sector at the national level (*Energieeinsparungsverodnung [EnEV]*) has changed the legislation towards renovation activities, making it impossible to contradict modernisation activities when the investor is able to demonstrate that they include a reduction of energy consumption. This is even the case with features that determine rent increases that go far beyond the total energy consumption of a household and determine only minimal reductions. In practice, this is often used as a 'door-opener' for modernisations that drive up the rents to levels that cannot be afforded by the majority of sitting tenants, so that these tenants move out and the flats can be let for a higher price again (for numerous examples, see http://mieterforum-pankow.net/?page_id=515). A Master's thesis that empirically studied 'energy efficiency renovations' in Berlin-Pankow thus demonstrated that:

> . . . energy-efficiency renovations have been used as a strategy for displacement. Energy-efficiency modernisations don't work to the benefit of the tenants, but they are used to shift their costs upon the tenants. This is also demonstrated by the transformation of tenant's community at 4 Pestalozzi Street over a period of three years which resulted in the displacement of half of the residents. For the tenants affected in the North-East of Berlin, energy-efficiency modernisation symbolises gentrification.
>
> (Schiebe 2016, p. 4)

Between Deregulation and Re-regulation

Since around 2015, the situation has been about to change again, and the consequence is a highly uneven landscape of deregulation and re-regulation.

Thus, on the one hand, after more than 20 years, all existing Urban Renewal Areas were abolished between 2009 and 2015. With this, a number of legal instruments that can be used to influence the development of these areas, as well as personnel mandated to control the realisation of publicly declared 'goals of urban renewal' (*Sanierungsziele*) stopped being available. As discussed above, the borough tried to absorb this development by declaring new Milieux Protection Areas in the areas covered by the former Urban Renewal Areas – yet in summary, the legal and technical opportunities provided by the latter are clearly superior to the first.

An even more problematic development unfolds with the forthcoming end of eligibility periods for the subsidies provided to support renovation activities in the 1990s. As in the section 'Rolling out the market, weakening public control', these covered up to one-fifth of the housing stock and were linked to rent caps and occupational rights for the borough. With some differences between the funding programmes, the commitment period agreed upon was between 15 and 33 years; the majority of subsidised houses will lose their status as a subsidised object, along with the resulting restrictions within the next years. Table 5.5 gives an overview of this development.

The segment of flats that have been subsidised and are affordable for low-income groups is, thus, about to shrink massively and will eventually disappear within the next decade (see ASUM 2012, p. 59).

On the basis of these facts, it seems as if the successive deregulation of the housing market experienced throughout the last decades would finally reach its end point and neighbourhood change in Prenzlauer Berg would eventually become a completely market-led affair.

At second glance, however, there are a number of new developments which provide a different picture. Throughout the last years, housing shortages and rent increases have become an issue that go far beyond Prenzlauer Berg, affecting wide parts of Berlin's population, as well as the residents of most other large German cities. This has fundamentally changed the political climate and, as a consequence, housing issues have become a top issue of both national and local political debate. In Berlin, this has also been supported by an enormous upswing of tenant movements and a new government coalition of Socialists, Social Democrats, and the Greens that came into power in 2016. On this basis, a number of reforms have been, or are about to be,

TABLE 5.5 Subsidised housing units and commitment periods in the Urban Renewal Areas in Prenzlauer Berg (Bezirksamt Pankow 2015, p. 311)

	Housing units subsidised	% of housing units subsidised	Commitment period	End of commitment
Kollwitzplatz	920	11. 0	20–33 years	2035
Winsstrasse	487	9.4	20–31 years	2034
Bötzowstrasse	396	11.1	20–28 years	2028
Teutoburger Platz	616	12.7	20–33 years	2036
Helmholtz-platz	2800	20.2	15–32 years	2038

implemented that have the potential to once again change the conditions for neighbourhood change in and above Prenzlauer Berg in considerable ways. The most important of these are the following:

- In March 2015, Berlin enacted a ban on the conversion of rental units into condominiums in Milieux Protection Areas; as a consequence, conversion rates have considerably slowed down.

- A number of boroughs have more actively started to use their 'right of first refusal' (*Vorkaufsrecht*) in Milieux Protection Areas and took over 59 properties between 2015 and mid-2019. Closely connected, the planning departments now demand prevention agreements (*Abwendungsvereinbarungen*) when properties are sold. These include an agreement to several conditions (e.g. not raising rents above the rental index, not making use of the right to claim personal use, etc.) in exchange for the boroughs not using their first refusal rights.

- The new government has ordered its six municipal housing companies to expand their stock from approximately 300 000 to 400 000 apartments currently. While the bulk of this expansion will be based on intensifying new building activities, the new strategy explicitly includes the acquisition of properties in high-price areas. Consequently, recent years have seen a number of takeovers of larger stocks, often former social housing, by municipal companies. As municipal companies are obliged to let their properties at comparably low prices, an expansion of the public stock is equivalent to expanding the supply of affordable homes.

- Private developments need to provide at least 25% of the total living space built to low-income tenants at rates comparable to public housing.

- In 2020, Berlin's government even introduced a Berlin-wide rent cap that froze all rents way below market rates. Unsurprisingly, this initiative was very controversial and ferociously fought by landlords, developers and the conservative opposition. Eventually, this battle was lost on 16 April 2021 when the Federal Constitutional Court declared the rent cap void, arguing that rent laws would fall under the legislative competence of the federal government, not the state government.

Most of these developments have only been introduced very recently, they are often implemented in incomplete ways, and are – as the debate about the introduction of a rent cap shows – under serious political attack. As a consequence, it is not yet possible to evaluate their long-term effect. Yet, at the same time, it is obvious that considerable changes are under way. In Berlin, current policies widely depart from the deregulation and market orientation of urban renewal experienced in the last two and a half decades. Whether these have the capacity to radically change the course of urban development and impede or even revise gentrification remains to be seen. For Prenzlauer Berg, however, it seems that most of these changes come at a point too late in time and it is difficult to imagine a de-gentrification process in an already fairly gentrified neighbourhood.

Gentrification with Brakes?

Summarising the course of gentrification in Berlin's Prenzlauer Berg, two contradictory developments that are illuminating are discussed in this section.

On the one hand, in the course of two and a half decades, state intervention has been considerably downsized and the market gained more control. On this account, public intervention followed a path of neoliberalisation (Brenner and Theodore 2002) known from many other cities in the world. The more the state withdrew from regulative interventions in the market, the more gentrification gained ground. Under this perspective, what happened in Prenzlauer Berg is reminiscent of what Hackworth and Smith (2001) termed a 'third wave' of gentrification, characterised by an intensification of gentrification processes, an involvement of larger-scale developers, declining resistance and an active role of the state in support of upgrading and displacement, at a first glance.

At a second glance, the role of the state seems much more ambivalent here. If a complex back and forth of regulations with regard to housing provision, allocation and pricing is already typical for Germany, this is particularly the case for Prenzlauer Berg. Throughout the last decades, new regulations have frequently been introduced that have resulted in changing legal and economic constellations and that have tried to erect protective walls against rent increases and displacement. The whole situation can be characterised as an intensive arms race between investors and local administrations (but also between the state government and the boroughs), in which the latter can, however, only participate with limited resources. Nevertheless, the outcome was an impediment to gentrification and a fragmentation of the chances for implementing rent increases, which differed tremendously according to different sets of situations and moments in time. The actual rent paid by a household today, thus, not only depends on the market value of a flat, but on the moment when it was renovated, on the regulations that applied at that moment in time and on the question of when and within which regulatory context the resident moved in.

The political reason for this is the split competencies between different scales of the German state, none of which can completely define the regulatory context for urban renewal on its own. Luxury modernisation and the conversion of rentals to condominiums have, thus, been made possible through national legislation and subsidised by national funding programmes and tax subsidies, while they were counteracted by the introduction of a cap on rents and standard definitions from the side of the borough. The state of Berlin supported the introduction of caps on rent to some degree, but refrained from stopping condominium conversion for a long time, while at the same time providing extensive subsidies for Careful Urban Renewal. The borough banned luxury modernisation, but needed to accept displacement when it contributed to energy efficiency and so on. The regulatory landscape around gentrification is, thus, split and contradictory, and this complex setting allows space to manoeuvre for all participants. To make things more complicated, all these constellations have changed over time, depending on the political coalitions in charge. The actual shape of urban renewal policies is, therefore, hardly a straightforward model of either neoliberalism or the welfare state, but rather a shifting and difficult to oversee political landscape that is based on the German three-tier system of government. It provides manifold chances for contestation on all three levels.

In this environment, gentrification can only proceed within a dense and quickly changing corridor, working its way around existing institutional limitations. In this sense, the institutional environment of renewal determined four different gaps, which each made different investment strategies attractive. The relevance of these gaps changed over time and each included different relations between investors, state authorities, and residents.

After the fall of the wall and before the declaration of rent caps, most of Prenzlauer Berg appeared to be a classical rent gap situation. The potential ground rent was considerably above the actual ground rent and this gap could be closed through modernisation activities that would justify rent increases to an amount of 11% of the renovation costs. These rent increases would often be beyond the financial capacity of the sitting residents to pay, and would result in their replacement by new tenants that were able to pay market rates. However, with restitution pending, it took a while until this potentiality could be realised, and the outcome was a distortion of the locational logics of upgrading. Moreover, with both direct subsidies and tax deductions in place, a large part of the bill was covered by the taxpayer, not the tenant.

This changed successively with the introduction of rent caps in 1995. Here, the potential ground rent was limited to an administratively defined amount that was set in line with the financial capability of the majority of the residents, at least for a period of five years after renovation. As Holm

(2006) has shown, the administrative definition of rent maximums was still cushioned by opportunities for tax deductions, but these became less lucrative over time. More importantly, however, these regulations could not really be controlled for new contracts. As a consequence, rents differed immensely between those tenants who moved into a building after renovation and those who lived there before.

At around 2000, a loss of tax incentives, the abolition of public subsidy programmes, and growing knowledge about tactics on the side of the tenants made this model successively disappear. From that time, renovation activities were largely based on the conversion of rental units into owned units. These were not only attractive because owning a flat is supported by the German state, but, more than this, because they enabled the dismissal of existing tenancies for personal use, thus, clearing the way for new tenancies at higher rates. The new legislation on energy efficiency in the housing stock eased this transformation, as it removed some of the legal barriers protecting existing tenants, whereas the introduction of limits for equipment features in Milieux Protection Areas made it difficult to achieve the most exclusive standards in the existing stock, thus, building barriers to super-gentrification.

With the almost complete renovation of the existing stock, rent gaps switched geographically to vacant lots on which new building could take place. In most cases, these lots were public property, so that the exploitation of investment opportunities in the form of new build gentrification rested on the sale of public land. Based on Paul Watt (2009a, 2009b), this nexus could be termed a 'state-induced new building gap'.

Recently, the picture has been changing again. On the one hand, since 2015 the conversion of rentals to condominiums has become very difficult in the nine Milieux Protection Areas that make up most of Prenzlauer Berg. As a consequence, tenure gaps resting on the conversion of rental units to condominiums cannot be realised anymore. Moreover, as most buildings are already renovated and the chances of achieving luxury standards have also been restricted through new legislation, the leeway for further investment is limited to some degree anyway. Thus, today, most rent increases are achieved through new contracts, not modernisation. This slows down residential change to some degree, but makes rent increases a sort of automatism that can hardly be controlled by local administrations anymore. This is all the more important as the end of commitment periods for the 15% to 20% of the housing stock that was renovated with the help of public subsidies opens a deregulation gap that can now be exploited on the basis of rent increases and new tenancies.

Over the course of 25 years, the existence and operation of rent gaps, thus, included different logics that partly worked synchronically and partly simultaneously to produce a rather complex set of dynamics. In each of the resulting gaps, the relations of landlords, investors, state authorities and tenants were subject to different institutional configurations.

In summary, this facilitated an ongoing gentrification. Where household incomes were low and weighted equivalence incomes were at a level of 69% of the (already low) average of East Berlin in 1993 when the first Urban Renewal Areas were declared (see Holm 2006, p. 237), in 2012 they were about a quarter above the average of Berlin, with both above average shares of low and very high shares of top income groups (see ASUM 2012, p. 17). Without a doubt, the social composition of the population has changed towards a dominance of better-off groups, and this is mainly due to the immigration of households with above average incomes. By all definitions, this is gentrification. Nevertheless, the manifold regulations enacted throughout the last decades resulted in differing bargaining positions between tenants, landlords and local authorities, which find their reflection in residential patterns that differ on a micro-scale.

The Table 5.6 illustrates this. It shows the characteristics of different statistics for the Urban Renewal Area of Helmholtzplatz by the type of housing they inhabit.

As is evident, there are immense differences between houses that were renovated using public subsidies and those where this was not the case. For the latter, rents paid are considerably higher and so are the incomes of the inhabitants. Here equivalence incomes are 50% above the average in the subsidised stock, while the share of unemployed households (meaning households with at least one unemployed household member) is up to five times lower. As houses that were made with the help of public subsidies and those where renovations proceeded without such assistance stand cheek by jowl, the outcome is a very uneven pattern of neighbourhood change. In many cases, this unevenness is repeated on a micro-scale, and within one house, one can find enormous differences with regard to rents paid, social status and income even between similar flats. Thereby, residents with a long tenancy usually have lower incomes and lower rents, whereas both rise in an inverse relationship to the length of occupancy. This pattern reflects the comparably better protection of long-established tenants within German rent legislation, which has been reinforced by the manifold regulations in place in Prenzlauer Berg over time. While these were not capable of stopping gentrification in the long-run, in the short run they have provided a system of trenches that can be used by sitting residents to hold out and fight rising housing costs to some degree.

The picture is, thus, contradictory: while gentrification was made possible by German rent legislation and fuelled by the restitution of properties to their original owners and tax subsidies in Prenzlauer Berg, it was also slowed

TABLE 5.6 Characteristics of different segments of the housing sector in the Helmholtzplatz neighbourhood (Based on ASUM 2012, p. 41)

	Not renovated	Renovated with public subsidies	Renovated with private money/full	Renovated with private money/partial	Newly built	Individually owned/ Condominium	All housing units
Share	15%	20%	43%	18%	4%	29%	100%
Average rents (€/m2)	5.45	5.47	6.75	6.17	7.74	6.90	6.22
Net-household income (€)	1739	1725€	2200	2000€	1900	2000	2000
Equivalence income (€)	1600	1260	1885	1500	1335	1800	1600
Unemployed households	7%	21%	8%	6%	4%	–	11%

down by numerous local regulations aiming to protect tenants against rent increases. The outcome of this is a prolongation of, as well as a spatial dispersion of, gentrification.

Notes

1. This chapter builds on previous work to a large part. Segments of it have already been published in Bernt 2003, 2012, 2015, 2016b.

2. The first three of these go back to the work of Andrej Holm (see Holm 2006). To some degree, the phases overlap – they should, therefore, be understood as an analytical rather than a chronological ordering scheme.

3. Subsidies for renovations were by and large designed after the German model of social housing, i.e. as a mix of non-refundable grants and low-interest credits that had to be paid back over a period of up to 30 years. After this, the status of social housing is lost and the building returns to have a 'normal' market status.

4. It should be mentioned that the restoration of historical ownership structures was highly contested in 1990, and was only made part of the Unification Treaty after strong pressure from the West German negotiators. The two main arguments put forward were, first, that there was a moral duty to undo injustices caused by Nazi and communist dictatorships and, second, that a return to previous property structures would be the safest way to secure the trust of potential Investors.

CHAPTER 6

Splintered Gentrification: St Petersburg, Russia[1]

Studying gentrification in Russia's second biggest city is anything but an easy undertaking (see Bernt 2016b). The term gentrification has only very recently entered Russian vocabulary and in most situations, Russians would use it in a way similar to terms like improvement or beautification.[2] The academic literature on the topic is very sparse and if gentrification is used at all, mostly by professional planners and urban academics, it seldom comes together with the critical connotation that is one of its major trademarks in the West. Both empirical works and conceptual arguments connected with the term gentrification in the West are largely absent here.

At the same time, remarkable changes in the urban fabric of St Petersburg are under way. The rapid growth of new commercial and elite residential complexes is successively changing the urban landscape and has been the cause of much concern for citizens. The loss of historic landmarks has given rise to widespread media coverage, political activity and grassroots organising around preservation concerns. Upscale building projects destroying the 'authentic' atmosphere of historic St Petersburg are a regular topic in everyday conversation. The lifestyle and the aesthetics associated with gentrification in Berlin, London and New York are well-known in St Petersburg too and it won't be hard to find equivalents of the type of bars that are so apparent in places like New York's Williamsburg or Berlin's Prenzlauer Berg.

Notwithstanding these similarities, wealth and poverty can still be found in very close proximity in almost all parts of the city. Thirty years after the start of the transition, poor and rich people still tend to live cheek by jowl and neither gentrification nor its opposite, the formation of marginalised neighbourhoods, have yet become characteristic of any area of the city. In fact, the

The Commodification Gap: Gentrification and Public Policy in London, Berlin and St Petersburg, First Edition. Matthias Bernt.
© 2022 John Wiley & Sons Ltd. Published 2022 by John Wiley & Sons Ltd.

lack of Western patterns of residential segregation is the common diagnosis of basically all empirical research conducted on this subject in Russia. Among scholars there is, thus, consensus that social segregation in Russian cities has not yet 'taken the form of separate habitation' (Axënov 2011, p. 53) and 'the poor remain very much a part of the fabric of central city society, living, typically, cheek by jowl with the newly rich in the same building' (Bater 2006, p. 17), or that 'enclaves of social segregation still have a rather insular character . . . a strong dispersion of different social strata within the city is more characteristic' (Makhrova and Golubchikov 2012, own translation). Marked segregation at the scale of neighbourhoods or even housing blocks, is virtually absent in most of St Petersburg, and the patterns of sociospatial differentiation still seem to have more in common with the Soviet Union than with advanced liberal capitalism.

The problem, thus, goes beyond linguistic issues. As for now, gentrification has not yet become a major issue for most Russian inner cities.[3] If segregation has taken place in Russia, it has proceeded in the form of suburbanisation and the construction of new elite housing and gated communities rather than as a gentrification of existing residential neighbourhoods. In the urban cores, even in the most prestigious areas, the situation is marked by a close spatial coexistence of extremely rundown and Euro-style renovated apartments, and a cohabitation of New Russians, workers, artists, migrants, pensioners and other social groups is the norm. Even areas with a proper supply of the commercial infrastructure typical for gentrification (i.e. Western-style bars and restaurants, high-end boutiques, branded fashion stores, etc.) still have a very mixed residential population, and this picture has barely changed over time. Summing up, large areas of separate habitation for different social groups – as are common for most Western cities – are rather the exception than the rule in St Petersburg.

Interestingly, this state of affairs runs contrary to the expectations of many urban scholars at the start of the transition. In fact, the majority of research conducted in the 1990s was dominated by expectations of a fast alignment of formerly socialist countries with urban patterns known from Western capitalist democracies. The terminus of post-socialist transformations was beyond a doubt thought of as a catch-up with Western patterns of urban development. A paradigmatic example of this position is the widely cited prognosis of Ivan Szelenyi (1996) who forecast massive suburbanisation led by middle-class households, but also a gentrification of inner city neighbourhoods and a rapid decline of the estates built between the 1960s and 1980s (which he expected to become 'the slums of the 21st century'; Szelenyi 1996, p. 315). Twenty-five years after this prognosis, it is obvious that enormous rent gaps can be found in most inner cities of the former

Eastern Bloc and that there are steep gradient between real estate prices in the centres and the peripheries of most cities. Yet, gentrification is at best in its infancy, and the difference between potential and actually realised ground rents rather expresses a desire for inner city living than actual transformations on a broader scale. Moreover, where reinvestment is taking place, its political economy and the resulting spatial patterns differ sharply from its Western counterparts. Due to the specificities of the Russian housing system, large rent gaps exist, but they can only be closed in a rather fragmented and spatially dispersed way.

Is gentrification, thus, a concept that is just not applicable to Russian cities and should be buried in this context? In fact, as this chapter aims to show, the situation is a bit more complicated, and although gentrification does not really follow the patterns familiar from classic studies in the USA or the UK, there are a number of features that make it a worthwhile case, useful for highlighting dimensions of urban change that are not yet very present in urbanist debates in Russia.

This chapter describes the three major dynamics in which gentrification can be observed in the historic districts of St Petersburg, and the difficulties under which it can proceed: the regeneration of existing residential buildings, the piecemeal construction of elite new housing, and the dissolution of *kommunalki* apartments. As I will show, these dynamics can only proceed with great difficulty and in an impeded, spatially fragmented and temporally stretched, and hardly predictable form. Moreover, they are embedded into an environment of decommodified and state-controlled regeneration, which is described in the section 'State-run repair and renewal' in this the chapter.

Unpredictable Regeneration Schemes

As discussed in Chapter 3, the pathway of transformation from socialism to capitalism taken in Russia has severely restrained the effective functioning of housing and residential real estate markets. The difficulties experienced by real estate businesses that have reinvested in the built environment in today's St Petersburg are a telling example in this respect.

Thus, in practice, if a developer aims to invest in a residential building, well-located in the centre of St Petersburg that has an impressive historical architecture, they regularly have to deal with 20–70 households, whose specific conditions and housing preferences are immensely heterogeneous. Within the same building, they might meet owner-occupiers who can hardly make a living and rent out parts of their flat in order to receive additional incomes, private tenants who can easily be displaced when a deal is set with the absent owner of the flat, municipal tenants who need to be resettled by the city, and owner-occupiers who have just bought their flat and are highly in debt with mortgage payments, as well as better-off households who are waiting for a chance to improve their housing situation. Reaching agreements with all households is a Herculean task in these circumstances.[4] Thus, in the words of one real estate agent interviewed:

> It is very difficult to redevelop a building. First, because finding a compromise with everybody is very difficult. And there is always a 'last passenger'. In one case we had a guy who would stay in his flat for five years. They had already cut gas and electricity and there was no light and no water – but he didn't want to move out. We call this the 'last resident'. The last resident can demand what he wants. Even if everybody except him agrees – the last one can stop the whole process. And then this was it! The thread of having a 'last resident' makes the risk for an investment unpredictable, both in terms of time and money . . . And there is no support from the administration. From their point of view, this is the business of the investor and they won't interfere. Sometimes, when residents start complaining, they even put clauses into the permission [for redevelopment, MB] and everything gets even more complicated. So, the whole operation becomes unpredictable. Second, in terms of economics, it is very difficult. If you want to redevelop these old houses you need to buy a new apartment for every resident, and that is expensive. . . . When the process of decanting starts, everybody tries to get the maximum. 'I live in a *kommunalka*, but you should give me a complete new flat. And it has to be in the centre . . .' It works like this: You have a building with 10 municipal tenants and 70 owners. People want to stay in their neighbourhood, so you need to find 70–80 apartments you can suggest. And these should not be somewhere at the fringes, at a bedroom community. There are these old grannies who survived the siege [in World War II, MB], who won't move anywhere, but within their neighbourhood. For matters of principle. Even if you give them a three-bedroom apartment in exchange for a room in a *kommunalka*, they won't move out. Third, there are these spacious flats of 200 square metres. Nobody needs 200 square metres. So you have to change the whole ground plan – which again is expensive. There is no economic rationality in this.
>
> (Interview with a real estate agent, 14 April 2015)

As vividly described in this quote, renovating an existing building is anything but an easy business. Calculations made by economists have clearly demonstrated that the costs and risks associated with renovating and preserving an existing historical building are significantly higher than new residential construction (Butler et al. 1999), so private investment in the existing building stock can only be made profitable in very small segments of the market. In this context, relocation costs of sitting residents have been estimated to account for up to 50% of total project costs (Butler et al. 1999, p. 30), and the messiness of the situation from the point of view of a developer has been vividly described as follows:

> In cases where individual apartments have already been privatised or are occupied by persons having beneficial rights to privatise, developers or brokers negotiate directly with the residents to purchase the apartment. Where the apartments have not yet been privatised, in many cases the developer's offer to purchase provides the incentive for the residents to privatise and improve their living conditions by moving to another, non-communal unit. The issue of holdout tenants, demanding excessive relocation packages, is commonly encountered by developers, and occasionally the demands of one or a small group of holdout residents result in developers abandoning the project. A developer or broker may also receive by mayoral decree the right to relocate the municipal housing tenants, and in return acquire ownership of their units . . . conditioned on the successful relocation of all of the current occupants at the developer's expense.
>
> (Butler et al. 1999, p. 30)

Put differently, in the privatisation of flats, the sitting residents have been given remarkable bargaining powers, irrespective of their income and status. As a consequence, the costs of displacement are incompatible with Western cases, and the regeneration of a property is harder to achieve.

On top of these risks, the developers need to obtain numerous approvals and permits from various governmental agencies. The development of a historical building, thus, always requires the approval of the State Committee for Heritage Protection (GIOP) as well as the City's Property Management Committee (KUGI) and the Committee of City Planning and Architecture (KGA), which is only given based on a detailed agreement including all sorts of obligations with regard to the preservation, resettlement, design and technical requirements. As most urban land is owned by the state, the lion's share of development is only possible if the city grants long-term lease rights, i.e. if local authorities and investors can make a deal. Depending on 'prior experience, staff capabilities and political contacts of the firm' (Butler et al. 1999, p. 31), this process can take three to four years. Moreover, the inclusion of different departments opens the door to all kind of behind-the-scenes actions on the part of the administrative actors, resulting in dealmaking that does not only follow official guidelines and development directives, but is also linked to the shady worlds of connections, corruption and politics.

Under these conditions, whether the planning agreement, relocation and eventually redevelopment of a lot can be achieved very much depends on the actual situation, i.e. the particular needs of the residents, their degree of knowledge and legal training, the intervention of preservationists, competition and rivalry within the state apparatus, informal relationships with decision-makers in the bureaucracy, etc. These factors don't follow a geographic logic, and the likelihood of successful reinvestment (i.e. gentrification), therefore, not only depends on locational considerations regarding the preferences of potential purchasers, but also on the estimation of procedural risks made by the investor. This sets two different logics into place, which work at the same time but in different directions.

World Heritage vs. Gentrification

Given these circumstances, it comes as no surprise that developers prefer to invest in empty lots or buildings and aim to construct new buildings rather than renovate. Instead of renovations, the major share of building activities, thus, takes place in the form of complete redevelopment, either by converting housing units from residential to commercial use or by constructing elite housing on top of existing buildings. Thereby, newly constructed buildings are usually considerably higher than previous structures, their architecture makes intensive use of glass and metal facades, and often large-size glass boxes are added on their roofs (see Figure 6.1). As for now, these developments are, however, rather scattered across St Petersburg, and usually they include only one or two buildings.

But why is this the case? Why have investors not concentrated on the most prestigious locations and transformed these in the ways known from many inner cities across the world?

The answer, again, lies in the specific institutional frameworks under which the real estate business works in St Petersburg. Most important, in this context, is the framework of heritage protection.

Well known as one of the world's most beautiful cities, the historic centre of St Petersburg forms a unique visual landscape that attracts millions of visitors every year. Most of historic St Petersburg has been under heritage protection since 1966 and was declared a World Heritage Site in 1990 (Nikonov 2012). In effect, the existing urban fabric is protected by a framework of international, federal and city laws and guidelines that not only prohibit the demolition or reconstruction of listed architectural monuments, but also enact a number of guidelines regarding the overall ensemble. These restrictions apply to 3934.1 ha of urban land in the centre alone (numerous areas

FIGURE 6.1 New structure on top of a historic building at Vladimirskyi Prospekt, St Petersburg in 2015 (photograph: Matthias Bernt).

in the historical imperial suburbs are protected too) and cover practically all land between the classical residential quarters of Petrograd and Vassilyevski Island in the north and the Obvodnyi Canal in the south. It is estimated that up to 1 000 000 residents of the city live in areas under heritage protection.

Although heritage protection is, thus, central to urban development in St Petersburg in general, and the demolition of existing buildings is rarely possible in theory, in reality, heritage regulations don't stop redevelopment at all. It has, thus, been reported that more than 1300 architectural monuments, nominally under state protection, have either been demolished or were 'in the phase of active destruction' in 2012 (Binney 2012). Independent newspapers and non-governmental organisation (NGO)-run websites have collected hundreds of examples of buildings that have been destroyed despite heritage protection (for shocking examples see www.save-spb.ru and http://kanoner.com/net). Given that the total number of objects of cultural heritage (historical and cultural monuments) is 8464, including 4213 objects of federal importance (https://gov.spb.ru/en/passport/culture), it becomes clear that heritage protection in St Petersburg is not a barrier to redevelopment but rather its modus operandi.

Under the conditions in Russia, demolition and redevelopment of formally heritage protected buildings are anything but impossible, yet they include a number of highly complex and rather difficult-to-track interactions between different state agencies and developers that, in effect, make planning regulations permeable to business interests. Moreover, the interests of the city government are ambiguous with regard to heritage protection – on the one hand, politics in St Petersburg have a clear orientation towards competition and business and economic growth, and in general there is an open-door attitude to investors. This has led to close forms of cooperation between city officials and developers that are fairly similar to those known from the literature about 'growth machines' or 'urban regimes' in the USA (see Tykhanova and Kholova 2015). On the other hand, city officials clearly understand that the unique urban fabric of the city is an asset that needs to be kept intact. Under these conditions, balancing growth and heritage protection becomes an ongoing problem that is worked out case by case. Against this background, a number of mechanisms have been set in place that enable people to bypass heritage regulations and foster demolition and redevelopment. The most important ones can be identified as follows (Nikonov 2012):

Lacunae in heritage protection: The concept of lacunae (*lakuny*) or 'plot excluded from the protection area' (*uchastok uzklyuchenyj v granitsach obyedinennoij ochrannoij zonyj*) goes back to the late 1980s, when local authorities looked for a way to make new developments possible within the heritage protected inner city (such as the destruction of the old hotel Angleterre in order to build a new one). Even then, redevelopment plans were hotly debated and gave rise to the formation of civic groups of preservationists that have remained influential until today. In a nutshell, the concept of lacunae is a camouflage term for the suspension of heritage protection guidelines for a particular piece of land that is located within the protection zone. This usually occurs in order to enable an investment project within the heritage protection zone without formally violating the existing protection framework. Thus, if an agreement between a developer and the city is achieved, and an expert statement has declared 'non-impediment' to the existing protection guidelines, a lacunae is declared by the planning authority, so that infills and reconstruction are made possible in that location. This especially happens in green spaces (like parks, gardens, playgrounds, inner yards) in privatised public spaces (kindergartens, cultural centres, sports facilities), or in commercial structures that are seen as dispensable by the authorities. Among the developments implemented on this basis are numerous buildings that have seriously altered the existing urban fabric (for instance, the highly contraversial Montblanc building), and raised intensive public concern.

There is no comprehensive register of lacunae, as new lacunae are continuously produced by the planning authorities. Moreover, while local planning authorities can propose lacunae, these need to be confirmed by a federal agency where 'objects of federal importance' are reviewed, and this happens on a case-by-case basis. However, a list of proposed 'plots [to be] excluded from the protection area' produced by the city government in 2004 alone contained 283 proposals in total, and since then new lacunae have been added continuously (see Nikonov 2012, pp. 160–161).

Planning mistakes: A second, rather bizarre mechanism through which protectionist regulations are overrun is the serial production of planning mistakes.

Newspaper and webzines have extensively documented some of these planning mistakes, and the usual procedure with which these take place can be described as follows. Developers acquire a particular property and start building without the proper permissions. Instead of immediately intervening, planning authorities keep their eyes closed and assert that they only gain knowledge about what is going on after the construction works are close to being finished. Cases have been reported where this happened with 73-metre high buildings, clearly visible from a large distance. After this, a 'planning mistake' is recognised and the investor is fined for violating the law. The size of the fine is usually insignificant when compared with the immense profits that can be made with this kind of development.

On this basis, even without taking into account the above-discussed lacunae, more than 200 new buildings were constructed within the heritage protected zone between 1988 and 2011 (see Nikonov 2012, pp. 160–161). All of these buildings have been ex-post recognised as planning mistakes. It is estimated that one-third of all construction activities in the city are started without the necessary permissions (Nikonov 2012, p. 161), and Trumbull (2012) noted 364 cases of planning violations in the first half of 2008 alone. The problem here is, thus, not a lack of regulations, but weak enforcement and implementation.[5]

Damaged houses: A third way that allows new constructions to be developed despite heritage protections is usually discussed under the rubric of 'damaged houses' (*avariinyje doma*).

The background for this is the fact that a large number of the existing historical buildings in St Petersburg have profound damage. Here, Russian planning authorities distinguish between two categories: a building is counted as damaged (*avariinyj*) if 40% of its structure is dilapidated and as ruined (*vetkhyj*) if 60% or more of the structure is affected. At the time of writing, around 5800 residential building were considered by the city to be damaged or ruined.

When a building is rated as damaged or ruined, local authorities have the right and the obligation to relocate its inhabitants in order to prevent hazards and guarantee the physical integrity of the citizens (Housing Code of the Russian Federation, Article 57). The declaration as ruined or damaged, thus, sets a process in action that is comparable to compulsory purchase orders in the UK or eminent domain in the USA, in the course of which public authorities can take control of an estate without the current owner's consent in return for compensation. When this takes place, federal law states that relocation must start and local authorities are obliged to offer the relocated residents appropriate alternative living space at a size and quality comparable to the apartment originally inhabited. Within this framework, owner-occupiers are, however, considerably advantaged in comparison to (municipal) tenants, as the former need to be compensated at the market price for the property lost, while the latter receive access only to another municipal flat or room of a comparable size. Private tenants who rented their living space from owner-occupiers are not regarded at all. In practice, relocation nearly exclusively moves the affected residents to peripheral estates (because land is cheaper and more easily available there and there is a severe shortage of accessible municipal apartments in non-dilapidated conditions in the inner city).[6]

Currently, there is no detailed survey available that would shed more light on the actual conditions of relocation, or on the living conditions of those resettled. Interviewed experts, however, evaluate the experiences as mixed from the perspective of the residents. Thus, while there are clearly residents who could use the relocation process to improve their living conditions (i.e. by using the compensation for their property as a co-payment that helps them acquire a better apartment), there are also numerous stories of residents who have been displaced to very peripheral parts of the city, far away from their jobs and networks. At least in these cases, it seems to be appropriate to regard the procedures based on the declaration of *avarinost* as a system of state-controlled displacement.

In basically all cases, after resettling the residents, reconstruction (*rekonstruktsiia*) is set into place. The declaration of a residential building as damaged, thus, goes hand in hand with reinvestment, and in most cases the investor is obliged to meet at least a large part of the relocation costs. The classification of a house as damaged can, thus, be seen as a sort of a universal picklock that grants extraordinary rights to the city and the investor, enabling them to simultaneously ignore existing property rights and bypass strict heritage protection regulations.

Interestingly, there is no clear procedure for getting a building on the list of damaged objects. Theoretically, residents, public authorities and investors can initiate the process by pointing the city government to the miserable state of a particular building. In reality, residents hardly ever do this. In contrast, it is more often the case that investors initiate the process, while residents try to get their building off the list. Once the city government learns about the problems of a particular building, it orders a technical expert review in which the damages are detailed and the building is classified. Again, these expert conclusions are regularly challenged by the media and local activists, as they are often seen as predetermined and subject to fraud and corruption. Moreover, there is no public inventory of damaged buildings, which strengthens the impression that classifications are produced in accordance with redevelopment projects.

While lacunae, planning mistakes, and the classification of structures as damaged enable state actors and business interests to break-up an otherwise fairly blocked situation, on the one hand; on the other hand, the constellation in which this happens is very open to contestation. The reason for this is that all exceptions to existing planning guidelines need to be based on expert statements and require public consultations. While there are numerous examples where these procedures have been manipulated, the very fact that these are required for modifying existing heritage protections also enables well-organised groups of citizens to challenge administrative decisions with alternative expert statements, legal contestations and all sorts of public disclosures and protests. As heritage protection has become the major concern of an increasingly well-organised social movement in St Petersburg, this, in fact, happens very often. Protectionist groups have been fairly successful in recent years and have effectively managed to stop, delay or alter a number of redevelopment projects, both on a small scale and on a citywide level. Using a broad variety of tactics, preservationists have become a kind of a third power group in St Petersburg. Even where redevelopment projects have finally been realised despite their protests, this group's actions have caused a situation where, as an interview partner expressed it, 'doing a bigger project in the centre of St Petersburg is asking for trouble' (Interview with a consultant, 15 May 2015). Against this background, preservationist groups are taken so seriously by developers that at a roundtable organised by the city government in June 2015, a number of big development companies declared publicly that they thought about suspending development projects in the inner city in general. Pointing at a list of 14 major development projects that had recently been stopped due to protests, a developer stated that real estate business in St Petersburg's centre would cause 'only headaches, but not profits' (ibid.). While there is certainly an element of lobbying in this claim, there is definitely an element of truth in the statement too. Given the general insecurity of the overall economic situation and the high interest rates, real estate development is usually a very short-term business in Russia. Against this background, a capacity to halt projects is a serious threat to the overall calculation developers make before deciding to invest.

To summarise, it can be said that a mix of factors is at work that makes new building gentrification a difficult business in St Petersburg. The most important of these are the complicated ownership structures, heritage protection guidelines and resistance by grassroots initiatives and preservationists. At the same time, city officials and developers have managed to establish a number of ways to bypass these complications, which enable developments on a case-by-case basis. In effect this has made redevelopment possible, but in a very dispersed and not geographically concentrated form.

The Dissolution of *Kommunalki* Flats

As discussed in Chapter 3, the persistence of communal apartments (*kommunalki*) is a remaining heritage of Soviet housing policies in Russia. While in 1994 only 6.3% of the total Russian population lived in communal apartments, this figure was considerably higher in the historic centres of the country, with 12.5% of all citizen sPetersburg, with its large historic building stock, had a particularly large stock of this form of housing and is often referred to as the 'capital of communal apartments' in public speech. Even today, 25 years after the start of the transformation, the situation with communal apartments has remained a major issue for housing policies in the city. Currently, according to the Housing Committee of the St Petersburg government, there are still 92 910 communal apartments in the city, inhabited by 283 263 households (Housing Committee 2014, p. 10) http://gov.spb.ru/gov/otrasl/gilfond/ gosudarstvennaya-program-ma-sankt-peterburga-obespechenie-dostupnym-zhi [no longer available]). This resonates with the 2010 Federal Census, which reports that 447 284 people lived in communal apartments in 2010, making up 9.5% of the total population.[7] Around two-thirds of the communal apartments inhabited in St Petersburg were built before 1957, so the share of *kommunalka* residents is likely to be as high as 15–30% in the inner city boroughs (*rayons*), where most residential building were constructed before World War I.

It should be emphasised that living in a communal apartment did not include social marginalisation at all in the Soviet Union. Well-positioned directors, famous actors, professors, writers and even the current president of the Russian Federation, Putin, have all found themselves sharing a flat with strangers at some time in their life. Since the start of the transformation, the situation has changed, as those with a higher income tended to move to separate apartments. As a consequence, *kommunalki* successively became a housing choice mainly for those who, for different reasons, were not able to earn high incomes. Under the conditions of Russia, however, these could be academics, as well as pensioners, students, construction workers and other groups. In short, while *kommunalki* were downgraded with respect to incomes, in general, those who live in them account for a broad variety of social situations. Nowadays, the social mix to be found within communal apartments is even on the increase, as more and more rooms in *kommunalki* are rented out. Given the tight housing market choices in most Russian metropoles, this brings in even more variety, as all sorts of newcomers and temporary workers rent rooms in communal apartments before moving elsewhere.

There is still a lack of empirical studies on the actual situation in this part of the housing stock. What is clear from interviews and media coverage, nevertheless, is that living in a *kommunalka* includes a broad variety of social groups and, nowadays, four groups of inhabitants can be distinguished:

1. Permanent residents (and often their family members of up to two younger generations) who remained in their rooms for a long time. In media articles it is commonly assumed that these are to a large extent pensioners who were given a room when they were young and who now lack the money to move elsewhere, or fear isolation in a different environment.

2. Young professionals or newcomers who haven't yet earned enough money to buy a separate flat and can't move into a flat inherited by their parents. For these inhabitants, living in a communal apartment is often seen as a sort of waiting room, suitable until a permanent habitation can be found and afforded. Artist communities can be regarded as a subgroup of this category that uses the opportunities provided by communal apartments for combining artist co-production with flat sharing.

3. Tenants who rent from the owners of individual rooms. As a general rule, tenants have rather low incomes and rent their rooms only for short periods of time. As discussed, tenants have very weak legal protections and need to pay a larger share of their incomes for housing costs. Especially in the least popular segments of the *kommunalka* sector, apartments are disproportionally let to stigmatised and disadvantaged groups of the population. Often, this includes temporary workers from southern former Soviet republics like Armenia, Uzbekistan, Tadzhikistan, etc. who live in overcrowded conditions to save as much money as possible.

4. While there is no reliable data on the actual scope of the phenomenon, it is evident that many (both former and still existing) *kommunalki* have been transformed into commercial spaces and are used as inexpensive

mini hostels. Even single rooms are frequently rented out using Airbnb. In these cases, dwellers use the living space provided on a short-term basis and include a huge variety of people, ranging from professionals of all kinds visiting the city for a business trip, short-term students, tourists and new immigrants in need of a stepping-stone.

In the mid-1990s, the most pleasing communal apartments experienced a first wave of gentrification, as the attractive palaces of the tsarist aristocracy were bought up quickly and were transformed into elite housing or for commercial use by venturing investors. As with much of the predatory privatisation taking place after the collapse of the Soviet Union, this included all sorts of scams, and there is much talk of developers grabbing a room in exchange for a bottle of vodka, on the basis of fraudulent privatisation certificates, or even of hired killings. Once all of the rooms were bought up, the developers renovated the whole apartment and let or sold it to New Russians with a taste for aristocratic living and the resources necessary to afford it. Naturally, the extent of these practices is not documented. There is also no comprehensive empirical study that sheds more light on these developments. What is certain, however, is that the dissolution of communal apartments occurred with great speed in the mid-1990s. Whereas *kommunalki* made up 22.4% of the total housing stock of St Petersburg in 1994 (and around three-quarters of the stock in the inner city), in 1996 the proportion had decreased to 14% of the stock (Gerasimova 2000). As new construction activities were close to zero at this time, it can be estimated that around 120 000 communal flats were abolished and restructured in a few years. In general, due to the higher attractiveness of the stock, this process was concentrated very much on the centre. Thus, interviewed residents of a street that is located five-minutes from the Winter Palace (Malaya Konyushennayay ul.) reported that the share of communal apartments in their neighbourhood decreased from 'around 80% to 90% to close to zero' (Interview with residents, 20 May 2015).

While it seems evident, that the dissolution of communal apartments, the displacement of their residents and the sale of apartments to better-off residents occurred quickly in a number of prime locations in the centre, the overall social and geographic results of this process are, however, not very clear. This is for three reasons. The first is that while it can be meaningfully assumed that a large share of dissolved *kommunalki* were sold to individual owners and the original population displaced, it is also the case that *kommunalki* disappeared from the statistics because their rooms were privatised by the sitting residents without any change in the actual residency taking place. Most likely, this has been the case with smaller apartments. The second reason is that the dissolution of *kommunalki* took place flat by flat and depended on a number of factors including the agreement of all residents, the geographical patterns of this process were rather dispersed. Therefore, we can regularly still find communal apartments in the less-advantaged parts of a building (sidewing, first floor), even in cases where the better apartments are inhabited by very rich people. The third reason is that many *kommunalki* were not transformed into high-end residential use, but were converted to places of business. Especially in the very centre of the city, mini hotels have mushroomed, and it is obvious that many of these were former communal apartments. As the dissolution of *kommunalki* undisputedly focused on the most prestigious and easy-to-get apartments, the best locations disappeared quickly from the market and what remained is often marked by their comparably inconvenient location (both within the house and within the city), that they are situated in dilapidated houses, or inhabited by residents who do not wish to move for any reason. As a consequence, the interest of investors has considerably cooled down in the following years.

To summarise, while it is evident that a first wave of changes led to a fast conversion of a large number of communal apartments, and to their upgrading and sale to better-off residents, very little is known about the exact extent and the geographic patterns of this development.

With the aim of speeding up the process and to provide individual apartments to all citizens, the government of St Petersburg launched a new programme in 2007 that set about achieving the dissolution of 80% of the remaining 120 000 communal apartments by 2016. The programme had five variants (see http://obmencity.ru/state/195/):

- *Variant 1*: All inhabitants (i.e. owners of individual rooms) of a communal apartment sold their rooms to an individual buyer. In order to support this, they received a social payout (*sotsijalnaya vyplata*) from the city that enabled them to buy an alternative accommodation in the real estate market.

- *Variant 2*: An inhabitant of a room in a communal apartment used the social payout to acquire one or several rooms from other residents and converted the apartment into single owner-occupied flat.

- *Variant 3*: If a communal apartment included uninhabited rooms that were owned by the municipality (e.g. because the sitting resident had moved out or had died without privatising), households that lived in other rooms of the communal apartment had the right to buy these rooms from the municipality for a reduced price, equal to 70% of the market value.

- *Variant 4*: The municipality supported reallocations and apartment swaps, so individual households were enabled to live in separate apartments.

- *Variant 5*: The communal apartments were also dissolved with the participation of an investor. In this case, the investor paid 70% of the market value of the apartment (and/or provided equivalent alternative accommodation) and the city paid the rest.

As is easy to understand, the calculation of the social payout plays a central role here, as it both reduces the price for the purchaser and pushes up the buying power a relocated resident can exercise when looking for an alternative in the market. It is defined according to the following formula:

$$\text{Social payout} = (\text{housing consumption norm})$$
$$\times (\text{number of household members})$$
$$\times (\text{average price per m}^2) \times (30\%)$$

The two crucial points in this formula are the housing consumption norm and the price. The housing consumption norm is defined at 18 square metres per household member for more than one-person households and 33 square metres for singles. While this size seems to be appropriate for a single room in a *kommunalka*, it would be harder to find a separate apartment matching this requirement. This is especially true for the historic building stock. There, small apartments are next to non-existent. Moreover, the housing price accepted as the basis of the calculation is defined by federal law as the average housing price in the region (i.e. the federal subject), in this case the city of St Petersburg. As price differences are considerable within the city, a subsidy based on the average housing price hardly supports the purchase of a flat with an inner city location. In other words, while the payout reduces the amount of money to be invested by the purchaser, it hardly enables a household without additional resources to find alternative accommodation in the inner city. It, thus, disproportionally benefits middle-class households that can mobilise savings just below the market price and use the payout as a bonus that enables them to make it in the market or purchase a better apartment than planned. In addition, the programme is very lucrative for a number of big development companies, that have specialised in providing small flats at the periphery for a price fine-tuned to the requirements of the programme. In fact, a number of real estate companies have specialised in this segment and provide carefree packages for residents from dissolved communal apartments. Thereby, the alternatives provided are sometimes located as far as 30 km from the city centre, in compact panel buildings and high-rises with small rooms and little commercial infrastructure.

For the inhabitants of communal apartments, the programme is, therefore, a double-edged blessing. While on the one hand, it enables families to fulfil their dream of an individual apartment, on the other hand, it also displaces poor residents to the peripheries of the city. In effect, it operates as a tool for the social cleansing of the inner city, which benefits residents with a middle income, but provides few alternatives except relocation for the poor.

Interestingly, the programme has already proven to be fairly inefficient and has clearly fallen short of its own intentions. Until 2015, only 9008 communal flats were converted – which is far below the goal of 96 000 to be achieved by 2016 (http://www.gilkom-complex.ru/2014-07-25-15-36-12/2009-10-15-16-34-45?item=584 [no longer available]). Several reasons for this failure are discussed in the media:

1. Residents often bargain for a price that is more in line with the market price for inner city apartments than with the calculations on which the programme is based.

2. Many inhabitants prefer to stay where they are because they are afraid of losing their social networks, higher commuting costs and psychological alienation if they moved to the periphery.

3. Many owners rent out their rooms in communal flats either temporarily (to tourists in summer times when they can live at their datcha) or permanently (because they live elsewhere). For pensioners and other low-income groups, this is often an indispensable source of income that they would lose if they were to give up their property.

To summarise, while many residents of *kommunalki* wish to live in a separate apartment, the conditions under which they can achieve this dream are not very favourable for a large part of them. The problems discussed are especially profound as, according to law, *kommunalki* can be only dissolved with the mutual agreement of all tenants. Therefore, if even one resident out of 15 refuses to move or cannot participate in the procedure (e.g. because they are in a hospital, abroad or in jail), the whole dissolution process is stopped. Thus, according to data from the Housing Committee of St Petersburg, while 8880 households applied to participate in the municipal programme for the dissolution of communal apartments in 2014, only 880 households were successfully resettled. This is the reason for a new legislative initiative that allows the social payout and resettlement without the consent of all inhabitants   (http://www.gilkom-complex.ru/2009-10-15-07-54-19/2014-07-25-17-26-14?item=1366 [no longer available]; http://www.spbdnevnik.ru/news/2015-04-10/zhitelyam-kommunalok-peterburga-khotyat-dat-pravo-na-rasseleniey-bez-soglasiya-sosedey/). Whether this will solve the problem, however, is debatable, as it will remain difficult to sell a communal apartment as long as individual rooms are still inhabited.

To summarise, the policy towards dissolving communal apartments finds itself in a deadlock. On the one hand, there is a demand for dissolved *kommunalki* and many residents of communal apartments would welcome resettlement into separate flats. On other hand, the conditions under which the dissolution of *kommunalki* takes place are not favourable for many of its residents. Moreover, individual inhabitants can exercise the power of veto and effectively stop the whole process. In recent years, the situation has become even more complicated, as budget cuts have severely reduced the resources available to the city to support the whole process (http://www.fontanka.ru/2015/04/15/045; http://www.svoboda.org/content/article/26953390.html).

Should the dissolution of *kommunalki* and the resettlement of their residents to other places be categorised as gentrification? In general, the aim to provide each family with a separate apartment is widely accepted and enjoys great popularity among the citizens of St Petersburg. The way the programme works (or rather does not work), however, is market-led segregation, as dissolved *kommunalki* are usually sold for prices that make them inaccessible for low-income residents, thus creating exclusionary displacement. Moreover, the benefits the programme has to offer to tenants that are resettled are not very balanced between different income groups. Thus, better-off residents can use the support gained from public funds to co-finance the purchase of bigger and better housing in more favourable locations through their superior buying power, while the poor with too few resources to afford expensive homes are doomed to smaller and more peripherally located accommodation.

State-run Repair and Renewal

More important, in terms of pure numbers, than new building and the dissolution of communal apartments is what Russians refer to as 'major repair' (*kapremont*). Most historic buildings in St Petersburg face serious maintenance problems and damages, so the question of how repairs and renewal activities are financed, decided upon and organised has quite some weight.

Within this field, we can find striking differences from what we have seen in Western countries. Theoretically, as in the West, housing maintenance is the responsibility of the owners. The problem, however, is that the privatisation of flats in the 1990s was not accompanied by a handover of responsibilities with regard to the building. Thus, while all flats in a house can be privatised, the house itself can still belong to the state. It is only since the introduction of a Federal Housing Code in 2005 that homeowners' associations were made possible and given the joint responsibility for the management and maintenance of the common areas of the house. In practice, however, there are a number of peculiarities that distinguish these associations from those known in Western countries.

First, in most cases the municipality holds a more or less considerable share in the housing associations. The reason for this is that most buildings have not yet been completely privatised, so the municipality still owns apartments. As the city is also the owner of the utility companies and the planning authority, this leads to numerous conflicts and tensions.

Second, homeowners' associations face a number of serious legal restrictions with regard to the preparation and utilisation of their budget. The Housing Code of the Russian Federation provides detailed guidelines for the levels of maintenance and utility fees that are detailed by regional authorities in accordance with recommendations provided by the Federal Ministry for Regional Development. Thus, for most homeowners, the calculation of costs linked to their property is very much subject to administrative decisions and bureaucratic procedures that are beyond personal control, rather than a matter of their individual calculations. In general, however, the fees determined by administrative decisions are rather low. Thus, in 2014, the fee to be paid for maintenance was 2.50–3.00 roubles per square metre (€0.04–0.05). As a consequence of these circumstances, most property owners do not have an interest in actively managing their common property, and it is a notorious problem to find an executive who is willing to take responsibility for the common areas and their general maintenance and utilities.

Third, as a general rule, most homeowners' associations are seriously undercapitalised. Hardly any association has the means necessary to finance major repairs and state support is indispensable for maintaining and renovating the buildings commonly owned. This dependency is further strengthened by how difficult it is to acquire and securitise loans for the renovation of existing buildings in the mortgage market.

Major repairs are, thus, mostly made in accordance with Regional Programmes for Major Repairs (*Regionalnyije programmy kapitalnych remontov*) that are set up by the federal subjects of the Russian Federation (in the case of St Petersburg, the city government). These define the priorities for renovation for each individual building for the next 25 years. The programme for St Petersburg was set up in December 2014 with a horizon of up to the year 2037 (http://46.182.26.58/SpbGovSearch/Document/7101.pdf [no longer available]). It is 1502 pages long and has 21 927 addresses for which the specification, the order and the date of major repairs are defined on the basis of a technical evaluation of the state of deterioration of main elements of the building and subsequent needs for repair. These considerations are done for different elements (walls, pipes, roofs, etc.) separately, so that different parts of a building are usually repaired over a long period of time.

To illustrate this, an excerpt of the repair plan for four neighbouring historical buildings in the centre of St Petersburg is provided in Table 6.1:

TABLE 6.1 Planned renovations in St Petersburg (Housing Committee of St Petersburg 2014)

Address	Built in	Repair of . . .						
		roof	facade	heating	water	waste-water	electric lines	basement
No. 11, 10th-Krasnoarmejskaya ul., Corpus A	1873	2021–2023	2015–2017	2015–2017	2015–2017	2015–2017	–	2012–2023
No. 13, 10th-Krasnoarmejskaya ul., Corpus A	1863	2018–2020	2027–2029	2021–2013	–	2024–2026	2027–2029	–
No. 14, 10th-Krasnoarmejskaya ul., Corpus A	1908	2027–2029	–	–	–	–	2015–2017	2036–2038
No. 16a, 10th-Krasnoarmejskaya ul., Corpus A	1899	2015–2017	2018–2020	2015–2017	2015–2017	2018–2020	2018–2020	2017–2029

As one can see, major investments in existing buildings follow the logic of administrative plans, rather than commercial considerations. They proceed according to criteria set up by engineers and are completed over the course of 25 years. There is no spatial logic in these plans: renovations are not concentrated in a specific area, nor are they equally distributed across the city. The major criterion for reinvestment is the physical condition of different elements of each individual building, not considerations about profitability or urban planning strategies.

The lion's share of funding for these activities is provided by state and federal budgets. The actual distribution of costs between the state and the owners is decided upon by an operator commissioned by the local government. It depends on the reserves from maintenance fees paid in the subsequent years and the availability of state subsidies at any given time. In most cases, the contribution of the property owners is rather negligible and has been estimated to be as low as 5% of project costs.

There have been attempts within the planning department of the city to move towards a more localised and integrated approach towards regeneration – yet, until now, with only few results. As early as 1994, the city had, thus, started preparations for the Preservation and Development of the Historical Centre Initiative. This initiative was even made part of a federal programme for the development of historical architecture in 2001, but it ended in a big scandal when it turned out that US$11.5 million were spent on consultants, without the actual work getting started. It took until 2011 for a new initiative to be prepared, aimed at (temporarily) resettling 340 000 residents from the centre of the city, extensively reconstructing the buildings they inhabited and renewing infrastructure and facilities. The original costs for achieving this goal were estimated at 300 billion roubles to be spent over 10 years. After a few years of preparation – and before anything had been started – the estimated costs were increased to one trillion and then four trillion roubles (approximately €60 billion), and the duration of the programme was extended to 30 years. It was assumed that around one-third of this gigantic sum would be covered by federal money and the rest of the money was supposed to come from private investors and public means (http://www.gazeta.bn.ru/articles/2016/03/03/227822.html). When the plans became public in 2012 and the first two renewal areas were set up at Malaya Konyushennaya and Northern Kolomniya – New Holland (http://gov.spb.ru/gov/terr/reg_admiral/otdel-stroitelstva-i-zemle-polzovaniya/materialy-k-celevoj-programme [no longer available]), this led to intensive discussions, as many inhabitants feared displacement. What happened instead was, until now – nothing. Reacting to cuts in the federal budget, the programme was put on hold in 2015 and a legislative initiative aimed at gaining better handling of planning instruments was scrapped by the state

Duma (the Russian parliament) in early 2017. Despite this striking lack of implementation, immense resources have been spent on technical preparations, including cost estimates and financial plans for each building in the area. By now, the huge gap between intentions and implementation has become a source of frustration within the planning apparatus and hardly anybody expects serious progress to be achieved any time soon. An interviewee even reported that serious discussions to remove the programme from the political agenda were taking place within the political leadership of the city. To summarise, operative forms of public–private collaboration, as they are typical for urban renewal in many Western cities, have not yet been founded and the modus of renewal keeps being statist. At the same time, the city has neither the financial means nor the personal and legal capacities, as well as the expertise, necessary for tackling the complicated issues described. What results from this is a bureaucratic mode of *kapremont* with little spatial logic, which is hardly capable of stopping the physical decay of much of the built environment.

Pro and Contra Gentrification

Is gentrification a concept that should be applied to the situation in St Petersburg? Given the enormous differences between this case and the classical processes of gentrification, the answer to this question can hardly be a simple 'yes' or 'no'.

On the one hand, one can meaningfully argue that there are a number of gaps at work and that the potential ground rent that could be achieved through their closure is enormously higher than the one actually paid at most places in the city centre. On this basis, a description of the gentrification potential could point towards the following issues:

- A large share of the housing stock is privatised, but not used as a commodity. It is, as Zavisca (2008) described it, a 'frozen asset'. Defrosting this asset would set free an enormous valorisation potential – yet this is hard to achieve and can only be realised on a flat-by-flat basis in most cases.

- This issue is even more pronounced in the case of *kommunalki* apartments, which could theoretically be transformed into spacious residencies. Nevertheless, this form of habitation hasn't ceased to exist and its dissolution has proved to be particularly difficult to achieve.

- Land occupied by dilapidated buildings, green spaces and vacant lots could potentially be another investment opportunity, but only if regulations applying to it (most importantly historical preservation codes, but also zoning and infrastructural requirements) are abandoned or bypassed, and the building is demolished and replaced by a new one.

Despite enormous difficulties, all these gaps provide enormous potential for uplifting the capitalised ground rent, and they are in fact closed on many occasions. Many flats privatised by their residents at no cost have been sold at considerable gain, and there is a market both for rentals and for owner-occupation in which housing prices are comparable to those paid in Western Europe. The number of *kommunalki* apartments, which once were the most common tenure in the centre of the city, has been considerably reduced, and in most cases this has been accompanied by the displacement of the former residents. Moreover, and despite historical preservation, numerous new buildings have been constructed in the city centre and nearly all of them are used either as commercial spaces or for housing in the upper segments of the market. Given these instances, one could meaningfully argue that gentrification *is* proceeding in the city of St Petersburg. Under this perspective, using the concept of gentrification would be useful because it would add a perspective to the discussion that is not yet present in public discourses, which focus on preservation issues and the loss of 'authentic' atmosphere. Using the concept of gentrification in this environment would allow for the acknowledgement of the role of real estate capital in the process, highlight the socially unevenness it is based upon and facilitate connecting individual cases to broader changes.

On the other hand, it is questionable whether all these instances make gentrification a major driver of change in any specific quarter of the city. While gentrification has always been perceived as a process happening at the neighbourhood scale in the literature, instances of gentrification have not yet piled up at this scale in St Petersburg. They are splintered over a wide territory and happen at the scale of individual parcels or even flats, but are nowhere concentrated to a degree where they would be capable of transforming a whole neighbourhood. Applying a concept that is deeply tied to notions of neighbourhood change is a challenge in this situation.

Underlying this difficulty are, however, differences in the societal and institutional environment into which gentrification is embedded. In this sense, the difficulties in applying the concept of gentrification to St Petersburg point towards problems of the concept itself; the sheer share of housing that is not used as a commodity reveals the weakness of gentrification in dealing with the matter of non-commodification. This is especially important for the case of Russia, as the lion's share of flats is de jurae private property and could be used as an asset here (as I pointed out in Chapter 3), but in practice non-commodified forms of distribution and allocation (like inheritance, barter and state provision) prevail. As a consequence, a large share of the housing stock is simply not subject to market forces and gentrification is limited.

In addition, different forms of tenure, i.e. different constellations of property rights and bargaining positions are fragmented at the micro-scale, usually within a residential building, but also – in the case of *kommunalki* – even within one flat. The monadic idea of a near-monopoly control over land (which forms one of the cornerstones of the rent-gap theory) doesn't make much sense in this environment, and, as a consequence, the appropriateness of one of the pillars of the rent-gap theory is called into question. In contrast to near-monopoly control, what we see here is rather a 'tragedy of the anti-commons' (Heller 1998): there are so many owners holding particular rights, enabling them to block decisions to prevent development from occurring. A paradigmatic example of this are holdout residents, which may have a relatively standard ownership title for an individual flat or sometimes even a room in a *kommunalka* only, but are able to block redevelopment plans for a whole building. In such a situation, the privileges that usually come along with landownership are considerably weakened, and defining who is entitled and capable to realise ground rent extraction becomes difficult.

Third, the relations between state and capital in Russia follow other logics than in the West. In a persisting dual state environment, formal institutions and informal networks between factions of the state apparatus and individual business actors do not have enough power individually to implement any meaningful strategy that goes beyond individual spots. What this results in is an untransparent and quickly shifting landscape of power in which no balance is achieved and neither the state nor a business are able to realise their goals. The outcome regularly experienced in St Petersburg is a stalemate in which binding decisions are the exception and neither the state nor real estate capital can enforce its goals.

In sum, there are pros and cons for applying the concept of gentrification. If gentrification is equalled with the invention of a neighbourhood by urban 'pioneers', middle-class invasion and the complete remake of the territory, it has hardly taken place in St Petersburg. As Vorobyev and Campbell correctly write:

> There are no 'neglected' or 'inexpensive' neighborhoods that can be assimilated whole by bohemian 'pioneers' and then later by the middle or upper classes. Of course these classes would like to be able to do this. But how is such segregation possible in the relatively homogeneously populated center? Forcible and economic displacement of residents is practiced, but it yields little in the way of real spatial stratification.
>
> (Vorobyev and Campbell 2009, p. 55)

The consequences of this situation for many residents are severe, nevertheless. Three points deserve to be mentioned here. First, the stalemate between development and the existing ownership and protection structures results in a situation where many historic buildings are increasingly dilapidated and their residents are trapped in a miserable housing situation. Second, the high degree of bargaining power on the side of owner-occupiers and preservationists, together with widespread corruption and weak law enforcement, have stimulated a form of redevelopment that is regularly based on the violation of legal protections, often in very brutal and forceful ways. Third, while owner-occupiers have considerable power to defend their interests, private tenants are a totally non-regarded minority in this situation.

At the same time, gentrification has certainly taken place in a large number of converted communal apartments and it is also evident in locations where historic buildings and parks have been demolished and redeveloped for upmarket residential use. There is, however, a major difference to classical cases of gentrification; these developments are not concentrated in a certain area. Although gentrification takes place, it is dispersed across a wide territory, and the separate bits do not come together to form a consistent whole.

The major reason for this is that reinvestment in the building stock follows two different logics in St Petersburg that work at the same time, but not in the same direction. On the one hand, investment decisions, as anywhere, are based on an estimation of market potential and location. They follow considerations about geographic centrality, the image of a place, the potential for reuse, etc. On the other hand, procedural risks weigh particularly strong in St Petersburg. Thus, privatisation to the sitting tenants has given many residents a degree of bargaining power (irrespective of status and income) that has proven to be a major impediment to many reinvestment schemes. Moreover, heritage protection guidelines and the watchful eye of preservationists insisting on compliance of investment schemes have complicated the business of developers too.

While numerous ploys are used to override these impasses, the solutions found can only work on a case-by-case basis. Lacunae, planning mistakes, declarations of emergency conditions, etc. cannot be enacted for a whole territory. As a consequence, investment opportunities cannot be concentrated in a particular area, so the potential for a complete remake of a neighbourhood remains low.

Altogether, the specific patterns of gentrification in St Petersburg point to a different relationship between market agents and the state in Russia. Here, the traditional setup is neither the laissez-faire state known from the liberal variants of capitalism, nor the corporate state guaranteeing social order, but a hybrid in which the borders between capital and state are blurred. As a consequence, the dynamics of gentrification follow a different logic in St Petersburg.

Notes

1. Parts of this chapter have been published in Bernt 2016b.

2. To give an example, an interviewed resident explained to me that the pricing out and the displacement of poor residents happening in his house in the course of renovation activities and apartment sales were actually 'a bad example of gentrification, instead of a good one'.

3. Moscow, with its extreme real estate market, is an exception to this rule. Here, gentrification has taken place within the 'garden ring'. Ostazhenka, a historic working-class neighbourhood in close proximity to the Kremlin, has become a widely cited example of gentrification (see Badyina and Golubchikov 2005). However, Ostazhenka seems to be rather the exception than the rule, and similar processes have not yet been documented elsewhere.

4. It should be noted that this situation is not only typical for Russia, but micro-ownership structures have remained an impediment to gentrification in many post-Socialist countries (see for example Marcińczak et al. 2013)

5. Interestingly, the policy of city officials with regard to planning mistakes seems to be very flexible. While there are serious hints of the role fraud and corruption play, there is also a wide range of reactions to them, extending from non-intervention to a straight ex-post enforcement of guidelines (in some cases, investors were even forced to remove floors built without permission). However, there is no straightforward logic under which this is done, and the whole procedure is both open to informal decision-making and corruption, as well as political contestation and civic resistance.

6. A short analysis of administrative documents about the relocation of residents from 'damaged homes', thus, shows that while the majority of damaged buildings were located in central boroughs (Admiralteysky, Vyborgsky, Kolpinsky, Tsentralny, and Petrogradsky Rayon), buildings constructed for replacement were exclusively found in more outward locations at Primorsky, outer Kolpinsky, Pushkinskyi and Krasnogvardejsky Rayon (http://www.gilkom-complex.ru/2014-07-25-15-36-12/2009-10-20-16-59-49?item=637 [no longer available]).

7. According to the 2002 census, 10.6% of St Peterburg's population lived in communal apartments. The eight years between 2002 and 2010, thus, saw only a slight decrease.

CHAPTER 7

The Commodification Gap

Universality vs. Particularity Revisited

In the previous chapters, I have discussed the historical development of the relationship between public policies and gentrification in Berlin, London and St Petersburg and considered how these impacted on the various ways in which gentrification operated in these three cities. It is now time to take a step back and see what the material collected suggests for the conceptualisation of gentrification. Has gentrification proven to be a useful concept for analysing diverse forms of urban upgrading, or should it rather be abandoned? What are the potentials of the rent gap as an analytical device, and where does this device reach its limits? How can the proposed concept of a commodification gap assist in tackling the puzzle between universality and particularity in gentrification research? In this final chapter, I bring together the major lessons learned from the empirical data, and reflect on what they suggest for the conceptual problems outlined above.

I will start with the question of whether gentrification should still be regarded a useful concept. Is it too broad a theory to allow for any meaningful analysis of fundamentally unique places, or should it be seen as a universally applicable analytical device? In my view, neither of the two options hits the nail on the head.

On the one hand, gentrification is indeed a universal phenomenon. Its two core components – reinvestment and displacement – can be observed in all three cities. In all three cases studied, reinvestment of private capital into the existing residential stock has taken place; this reinvestment was determined by the search for profit and was based on a gap between the actual ground rent and the potential ground rent, which motivated investors to

The Commodification Gap: Gentrification and Public Policy in London, Berlin and St Petersburg,
First Edition. Matthias Bernt.
© 2022 John Wiley & Sons Ltd. Published 2022 by John Wiley & Sons Ltd.

uplift housing units and put the stock to its highest and best use. In all three cases, this led to a displacement of low-income households by better-off residents. By all definitions, this is gentrification.

Beyond categorisation, gentrification has also proven to be a useful concept for investigation, as it provides a perspective that allows one to connect the political economy of housing production and allocation with residential segregation. It brings issues to the table that easily get lost when the talk is simply about changing consumption patterns, loss of atmosphere and authenticity, or a declining social mix. Using the concept of gentrification is eminently political because it links changes in what Ruth Glass called 'the social character of a neighbourhood' (Glass 1964, p. xviii) to actual actors and interests.

At the same time, the appearance of gentrification is puzzlingly different in the three cases observed.

In Barnsbury, gentrification has for a long time been a process in which middle-class purchasers acquired houses for personal use as owner-occupiers, thereby displacing low-income tenants. Tenurial transformation, occupational change and displacement proceeded hand in hand. This was possible because the dynamics of gentrification were embedded into a state policy that supported homeownership through taxation, direct subsidies and the sale of council flats and was operated by speculators exploiting the value gap. In the new millennium, we face a completely different landscape: the traditional value gap has disappeared and gentrification is mostly driven by an inflation of house prices, which continuously increases the potential ground rent, even without previous devalorisation or tenurial change. As gentrification has become more integrated into global financial circuits, it has developed into super-gentrification, by which even the original gentrifiers and other dwellers with their socioeconomic profile are driven out. Together with this, private renting has returned and the established association of gentrifiers with middle-class owner-occupiers, which has been central for the British discourse about gentrification, is being blurred. Gentrification has mutated and there is a sea change between what gentrification used to be in the times of Ruth Glass and what it stands for in today's London.

Prenzlauer Berg is both similar and different to this picture. On the surface, some of the changes experienced in London can be observed here too. As in London, private investment was channelled into under-priced homes in an under-valorised location. The outcome was rising housing costs, leading to a displacement of poorer segments of the population and, even more importantly, excluding other poor people from having a chance to still move into this area. As in the UK, private investment strategies have been supported by the state, both with direct subsidies and with tax giveaways in Berlin. Yet, there are also particularities that make Prenzlauer Berg a case quite different from Barnsbury. First, the changes described above were crucially effectuated by the decision to 'restitute' properties to their original owners. Clearly defined individual property rights, as they are for the most part an unchallenged premise of gentrification theories, were hardly in existence when the Berlin Wall came down. On the contrary, they were only established through legislation that combined privatisation and commodification to a degree unknown before, even in the UK. Second, gentrification has not been connected to tenure polarisation, as was the case in the UK. Reflecting the shape of the German housing system as a 'unitary rental market' (Kemeny 1995), gentrification has almost completely taken place within the rental market, and even where a conversion of rental flats into individual ownership has taken place, this was mostly used to enforce new tenancies. The major reason for this is the different rent legislation in Germany, which protects renting to a much higher degree than is seen in the UK. The combination of a dominance of private renting with comparatively high protections for tenants does, however, not prevent 'bypasses' through which investors can achieve upgrading for a satisfactory return on investment. Nevertheless, it effectuates in numerous defensive positions in which tenants can take cover for a while, thus delaying and fragmenting displacement and population change. Applying a metaphor used by Gramsci, the way gentrification proceeds in Berlin, is a 'war of positions' rather than a 'war of manoeuvre', (Gramsci 1971, p. 238ff.) and the consequence is that until today, low-income households that have lived in the neighbourhood for a long time can be found living cheek by jowl with increasingly wealthy incomers who pay considerably higher rents. While gentrification was, thus, made possible through restitution tax incentives, and the successive withdrawal of the state, it was at the same time regulated through a complex system of tenants' rights and public interventions that limited the realisation of highest and best use of property. The effect has been an arms race between investors and neoliberal politicians (mostly at the state level) who were pushing for deregulation, on the one side, and tenant groups, progressive politicians borough administrations, planning offices and welfare organisations attempting to curtail gentrification, on the other. The outcome was a ping-pong of regulation, deregulation, and re-regulation. The socio-spatial consequences of this constellation are remarkable differences between the residents of the neighbourhood on a micro-scale with regard to length of tenancy, regulation status of the flat affected, income and tenure.

In St Petersburg, progressing gentrification has been even more difficult. The privatisation of housing introduced with the start of the transition has only worked ineffectively in bringing to the fore free, self-regulating markets, and even today, three decades after its start, is not finished. The outcome has been a splintering of property rights on a micro-scale, generating enormous barriers to investment. In such an environment, to assume that property owners would exercise monopoly rights that they use to maximise the returns they can draw from a piece of land, would be a grotesque mischaracterisation of the situation. Even in the most attractive areas, only parts of the stock are held as commodities, whereas the rest is to some degree a frozen asset that is not, or only with restrictions, traded in the market. Reinvestment is also made difficult by the insecure financial environment and by inefficient and at times corrupt planning authorities and regulatory practices. The excess weight of all these factors has led to a splintering of gentrification, instances of which can be found over a wide territory but fail to aggregate at the neighbourhood level. As gentrification is usually regarded as a process of neighbourhood change, this implies serious challenges for the gentrification theory.

To summarise, gentrification follows very different dynamics in the three cases examined, resulting in different temporalities, spatial patterns and political issues connected to this form of urban change. Gentrification as a process is not only essentially different between the three cities, but also between different neighbourhoods in each city, and even within one neighbourhood at different times and with regard to different pieces of land in it. Instead of one gentrification with a capital G, one should, thus, speak of a multiplicity of gentrifications.

The various manifestations of gentrification are but a symptom for different ways in which real estate and housing markets are 'embedded' in London, Berlin and St Petersburg. Generations of struggles have led to very dissimilar 'relationships of forces' (Poulantzas 2000 [1978]) in the three cities, which have 'condensated' into diverse, contested and changing regulatory landscapes. This results in dissimilar assemblages of institutions, organisations, legal frameworks and property rights in the three cities – which have selectively privileged some interests over others (e.g. owner-occupiers over private landlords, restituents over residents and flat owners over developers) at a specific point in time. The outcome is that we see different ways in which markets are built, capital is invested and displacement is achieved. In this work, I have suggested the new concept of commodification gaps as a tool that enables us to identify the interrelations of these arrangements with gentrification in a holistic way. Describing the interplay of market forces and political regulations, I have suggested 12 commodification gaps (see Table 3.6) and applied them to develop an in-depth understanding of the histories of gentrification in three neighbourhoods across Europe. The following gaps were identified.

Early gentrification in Barnsbury was accompanied by a value gap between the tenanted value of a flat and its value when sold for owner-occupation. When the Right to Buy Policy was established by a Conservative government in 1980, this value gap became accompanied by a privatisation gap in which the accumulated potential piled up in council housing was unfrozen and transformed into a mechanism that ensured a continued supply of gentrifiable flats in an already gentrified area. With the new millennium and the introduction of new financial instruments, the potential gains to be made from under-valorised properties became more and more embedded into global asset strategies, thus transforming the formation of potential ground rent and facilitating the emergence of both the inflationary gap and the derivative gap. Finally, the deregulation of private renting assisted in restoring profitability in this sector and led to a rental deregulation gap, which at least to some degree replaced the traditional value gap resting on owner-occupation.

In Prenzlauer Berg, in general, gentrification rested on the restoration of private property after the fall of state socialism (which, together with radical privatisation, was a central piece of conservative transformation policies in the housing sector of the former German Democratic Republic). Once it was set in place, four gaps emerged that proceeded simultaneously but with different strength at different moments in time. In the 1990s, the most important, in this respect, was the modernisation gap, i.e. the opportunity provided by German legislation to increase rent by 11% (9% since 2018) of the modernisation costs annually. Given the state of under-maintenance, modernising the housing stock was crucial. However, closing this gap would have been very difficult without the generous opportunities for tax deductions provided by the federal government until 1998, which enabled financing modernisation through tax savings. At the same time, the borough government introduced new regulations, and rents stemming from modernisation were capped at an affordable level for a couple of years. The more the modernisation gap was closed, the more important became two other gaps that substituted the regulatory limits set on modernisations, but were also based on these. Thus, when a higher standard would justify higher prices to be gained in the free market, a strong motivation was set in place to abolish existing tenancies (in which rent increases are limited) and to rent out to new tenants. I have called this the 'new tenancy gap'. In addition, the conversion to condominiums became more important, both as a consequence of taxation policies and because it enabled the owner to claim personal use, thus enabling the abolition of existing tenancies and achieving new ones at a higher price. In the German system of rent regulations, these three gaps are closely interrelated and much of the operation of gentrification in Prenzlauer Berg was based on switches between them. After the turn of the millennium, however, these three gaps were either closed or their closure was made difficult by new regulations (e.g. the ban of luxury modernisations in Milieux Protection Areas). As a consequence, capital switched to lots that were not built-up and gentrification proceeded in the form of new built gentrification. Most recently, a deregulation gap is opening up: as the commitment period of subsidies used to support renovation activities in the 1990s expires, the buildings affected are re-subjected to market forces and all three gaps described are opened. If and how this gap can be exploited is, however, again unclear as new rent legislations are being introduced (and challenged) in Berlin that have the potential to severely complicate gentrification.

In St Petersburg, one can find gaps between the capitalised and the potential ground rent that are caused by the price differentials between different tenancies; gaps induced by the opportunities to bypass or violate preservation restrictions, demolish existing structures and build new 'Euro standard' houses; and gaps resulting from the dissolution of *kommunalki* apartments. Together, these gaps reflect the unsettled and not yet finished transformation of the Russian economy from socialism to capitalism. All the gaps described have been and continue to be exploited in one or the other way – but none of them has been very effective. Remaining decommodification and strong bargaining rights have seriously impeded the progress of gentrification and led to slowing it down and dispersing its occurrence over a wide geographic area.

Across three cities and several decades,[1] the 12 gaps identified have proven to be a useful instrument for uncovering how diverse institutional environments have shaped real estate and housing markets in a way that has brought about different gentrifications. The gaps identified reflect these arrangements and enable a deeper understanding of the dynamics of upgrading at specific times and places.

From an individualise perspective, this could easily be regarded as an argument for the incomparability of different contexts and proof that gentrification as a concept loses its explanatory value when applied outside its place of origin. It could, thus, be argued that, for example, the closure of the value gap in London in the 1970s has nothing to do with, say, the dissolution of *kommunalki* flats in St Petersburg in the 1990s. One could refer to the fact that the first rested on the support of homeownership by British governments, whereas the latter went on without much state support. Moreover, whereas the conversion of rental flats occupied by low-income households and their sale for owner-occupation could only proceed with the exertion of immense pressure on tenants, *kommunalki* were such an unpopular form of habitation that many residents happily put it behind them as soon as they had a chance to do so. One could also point to the spatially concentrated character of gentrification in Barnsbury and contrast it with the splintered form of gentrification in St Petersburg or set the increasinlgy elite composition of Barnsbury's population in contrast with the remaining social mix in Russia's northern capital, etc. In a similar vein, one could proceed with

Prenzlauer Berg, or even distinguish between different periods in one area. In short, arguments supporting the claim that gentrification is too small an umbrella for covering different cases of urban upgrading are easy to find.

The same could, however, also hold true for a universalist perspective. Here, one can point to the fact that capital was eventually reinvested into undervalorised stocks in all three cities. It is also easy to see how the reinvestment of capital went hand in hand with increasing housing costs and a displacement of poor residents, thus implying an 'embourgeoisement' of the areas affected. From this perspective, the differences between St Petersburg, London and Berlin are rather a matter of degree than of kind, and Barnsbury could be seen as a frontrunner of a process in which Berlin and St Petersburg lagged behind.

Thus, in a way, both perspectives work well. At the same time, however, both fail to make sense of the differences observed and do very little to advance the conceptualisation of gentrification. Applying an individualist perspective easily slides into neglecting a fundamental nexus that holds true across all three cases: capital is invested into land and housing for the sake of profit, which can only be achieved by increasing the housing costs, and this results in the tendency towards the exclusion of poorer segments of the population. Without acknowledging this nexus, local studies of gentrification remain pure description, incapable of integrating differences into wider explanations and failing to produce knowledge beyond the case. Applying a universalist perspective, in contrast, more often than not comes together with an ignorance towards individual specificities and with a relegation of these to the status of contextual factors. In this perspective, the differences between the closure of the value gap in Barnsbury in the 1970s and the dissolution of *kommunalki* flats in St Petersburg after the fall of communism, to stick with the example, are isolated as the specificities of each case, not taken as a starting point for developing a deeper understanding.

How then can the relationship between the universal and the individual in gentrification be summarised up in a more abstract way? To do this, it is necessary to determine the relationship of factors that are unique to each case studied here to a general common (gentrification). Thereby, we must abstract from the particular conditions of each case and separate the elements that are part of a common relation. We must then examine whether the relations between these particular elements to the totality are internal and necessary or external and contingent. We, thus, need to ask whether the particular mechanism – the value gap in the UK, the new tenancy gap in Germany, the dissolution of *kommunalka*, etc. – are necessary for producing gentrification, or not. Can there be gentrification without these gaps and vice versa?

The answer is complicated. Clearly, gentrification can proceed at one place without including the conditions necessary to produce it elsewhere. A change from renting to owner-occupation was crucial for the operation of the value gap in London in the 1960s–1980s, but it hasn't played much of a role in Germany. Neither has the dissolution of *kommunalki* flats been of any importance to Berlin or London, as this form of tenure didn't exist. In St Petersburg and London, in turn, modernisation gaps – which stood in the centre for the operation of gentrification in Prenzlauer Berg – had not much relevance because rent legislations are different. At the same time, without modernising flats, or without converting rental flats into owner-occupation, or dissolving *kommunalki* flats, gentrification would have remained only a theoretical possibility in Prenzlauer Berg, Barnsbury and the centre of St Petersburg, and failed to become a reality. The local gaps are, thus, necessary for gentrification to happen at a specific place and time, but of no relevance elsewhere.

In turn, the existence of these gaps is only possible where there is a gap between potential ground rent and capitalised ground rent: if rents were already at their highest and best use level, no investor would be set on burning money by uplifting the quality of a housing unit at a particular piece of land by installing a modern heating system, buying out sitting tenants, vacating the apartment for owner-occupiers yet to come, etc. Local gaps and the potentiality of gentrification are interrelated, and they mutually depend on each other for their existence.

The issue at stake here is multiple causation: gentrification is unlikely to happen because there is a theoretical difference between a capitalised and a potentially higher ground rent alone. Making this rent gap work demands other gaps that are formed by the actual institutional context in which a rent gap operates. Gentrification, thus, results from the combined impact of this general gap with multiple local gaps, and this interplay finds its expression in numerous individual gentrification*s*.

Figure 7.1 sketches the interplay of universal and particular factors in bringing about gentrification. Distinguishing between gentrification as a universal phenomenon (left side of the figure) and diverse individual gentrifications (right side of the figure), it describes how gentrification is simultaneously caused by general conditions that need to be met by all cases and particular conditions that only work at a specific time and place. Thus, while the existence of land rent, uneven development and unequal income distribution is endemic to capitalism and, therefore, a universal feature, these general conditions need to be complemented by specific individual mechanisms necessary for bringing about the actual empirical result (Gentrification Type A–D). With regard to gentrification in general, these mechanisms are contingent, with regard to the individual empirical result they are necessary. As land rent and capital accumulation are nowhere to be found outside specific institutional contexts, there is a necessary nexus between the universal causation of

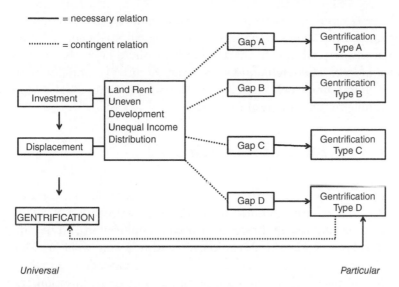

FIGURE 7.1 The causation of gentrification as both universal and particular.

gentrification and particular sets of conditions that need to be put to work, In order to allow gentrification to happen. While rent gaps may exist in all places from A to D, there need to be particular gaps in gentrification A–D, in order to allow the rent gap to be realised. As a consequence, gentrification can only proceed within specific institutional environments that bring about specific mechanisms that capital needs to work its way around.

In a nutshell, this understanding suggests the three conclusions explored below.

First, each gentrification is unique. Gentrification can proceed in place A through gap A (e.g. the classical value gap), whereas it proceeds in place B through gap B (e.g. the dissolution of *kommunalki*). As the two express different relations between state, capital, property and people, they are different formations.

Second, without the general mechanisms expressed in the rent-gap theory neither gap A nor gap B can exist. The exploitation of a differential between the tenanted value and the vacant possession value of a flat upon which gentrification rested in London in the 1970s is only possible where landed property exists, capital switches are possible and unequal income distribution leads to differential buying powers. If any of these conditions fails to exist, it is all but impossible for the value gap to emerge. The universal conditions for gentrification are, therefore, simultaneously preserved and transcended in the individual gaps.

Third, it is, thus, crucial to see gentrification as overdetermined. Disentangling gentrification as a universal phenomenon from the conjunctures of local circumstances, which are reflected in the manifold individual gaps, is just about impossible. Generalised rent gaps and localised commodification gaps are therefore Siamese twins – neither of the two can cease to exist without extinguishing the conditions of the other.

Gentrification and Decommodification

While the particularities expressed in local gaps are specific to each location studied here, they also have a common logic. This logic is decommodification.

It needs to be emphasised, however, that commodification and decommodification stand in a dialectical relationship and it is only when housing becomes a liquid asset that its potential for capital accumulation can be exploited to its full potential. Commodification and decommodification are, thus, essentially interlinked.

The actual form of this nexus is, however, organised differently in the diverse historical and institutional contexts. There is nothing uniform about the way in which housing provision is embedded into or disembedded from the society, and different societies have found very various arrangements for the production, allocation and consumption of housing. The reason for this variety rests upon the history in each location, i.e. the existence of different political and cultural traditions, class compositions, different political alliances and compromises, and other factors that differ between countries, regions and cities. How, where and by whom housing is provided, as well as the way housing is sheltered against market forces is an outcome of struggles – and these are time and place specific. Put differently, housing decommodification is not coincidental, but stems from political contestation.

At the same time, the balance between commodification and decommodification developed in a particular context is not a zero-sum game between the two; the way housing is organised in a particular society defines the bargaining positions of households, landlords, investors and public officials. The differentiated protection of private tenancies between the UK and Russia, on the one hand, and Germany, on the other, or the different treatment of social housing as residual (UK), integrated into the market (Germany) or leftover from failed privatisation efforts (Russia) provide vivid examples of this. The actual shape of a rental contract, how access to social housing is provided, how mortgages can be accessed and other issues define the rules of the game for a wide spectrum of actors involved in the production and consumption of housing.

While decommodification is central to how a housing market is built, it should not be misunderstood as 'more regulation' (and commodification as 'less regulation'). As the British example shows, commodification required the strong hand of the state to proceed, and both the introduction of new financial instruments and the commercialisation of the council stock would not have been possible without generations of neoliberal policy initiatives. In post-socialist Russia and East Germany, this is even more visible. Both the restitution of properties to their original owners (or their heirs) and voucher privatisation cannot be mistaken as deregulation. Quite the contrary, they were acts of decisive re-regulation in the interests of particular social groups and connected to specific political strategies.

A second word of caution is necessary with regard to the relationship between commodification and decommodification. Decommodification hardly ever goes together with the complete eradication of the commodity status of housing; 'it is not an issue of all or nothing' (Esping-Andersen 1990, p. 37); rather, there are very different forms in which housing can be decommodified. Thus, as in the case of British council housing, parts of the housing stock can be completely taken out of the market and managed in ways that have nothing to do with the interplay of supply and demand. Alternatively, as in Germany, quality standards, land use, sales prices and rents can be regulated in such a way that the conditions under which a home is produced and consumed are co-determined by market conditions and administrative procedures. As the case of Russia shows, decommodification can also be an outcome of market failure, where houses and flats are individually owned assets in theory, but only exchanged with great difficulty in practice. The point here is that there are not only various ways in which decommodification can appear, but all these ways can include different degrees of decommodification, and these are an outcome of the 'relation of forces' (Poulantzas 1978 [2000], see the section 'Universality vs. particularity revisited') at different spatial scales, which can be combined in numerous ways. In short, decommodification is not a substantive issue, but a matter of how, where and for whom.

Meeting the Challenge: New Directions for Research and Politics

It will be clear that this argument is directed against the idea that it is primarily economic forces that drive urban change.

Critical gentrification scholars tend to see the state as a mere instrument of the ruling class, used to extend the working of market forces. Political determinations of gentrification are only of interest in this view when the state acted as an active supporter of gentrification. Such conceptions are, however, likely to be myopic towards contradictions, alternatives and differences. Treating the state as subordinate to capital is excessively narrow (to borrow a phrase used by Smith 1977). A broader theory of gentrification must take the role of policies, polities and politics seriously.

With the novel concept of the commodification gap, I suggest a conceptual alternative. It is meant as a heuristic device capable of addressing the nexus between commodification and decommodification, providing scholars of gentrification with a tool to study the relationship between the operation of gentrification and the institutions underpinning, modifying and limiting it.

In this respect, the commodification gap builds on the rent gap, but focuses on the embeddedness of markets. Its usefulness lies in the fact that it enables one to understand how gentrification is brought about by differentiated regulatory landscapes that themselves are the outcome of political contestation. This, I argue, opens up a perspective that is capable of bringing about an understanding of the political foundations of gentrification. It, thus, adds a conceptual device to the toolbox of gentrification studies that connects theoretical arguments developed in the field of economic sociology with Marxist theories of uneven development in a way that is applicable to observed gentrifications and enables the researcher to understand their systemic features in tandem with the actual struggles they are based upon.

In constructing this device, I have leaned on empirical research that was inspired, shaped and limited by the three cities I have studied. Typical for them is a predominance of formal housing and real estate markets. Moreover, I have focused on residential change, leaving commercial gentrification aside. Arguably, this has led to less attention being paid to changes in infrastructure provision, matters of public space, etc. Nevertheless, I hope that other researchers will find the commodification gap useful enough to apply it to other contexts to test its validity and possibly modify and change this concept. In my view, three areas appear to be particularly relevant in this respect.

First, as I have already noted, I have only investigated the operation of the commodification gap in contexts where private property and formal housing markets are the norm. However, this is not the case in large parts of the world, but this doesn't necessarily make the argument inapplicable. Non-market forms of land allocation can form persistent barriers to commodification. Persistent decommodification or non-commodification of large parts of the housing stock can, thus, lead to 'real estate frontiers' (Gillespie 2020) in which there is a clear demarcation between market housing and other tenures. Moreover, the nexus between commodification and decommodification doesn't necessarily need to be negotiated through the state, but can also be embedded into property rights that are splintered and negotiated through community, caste or kinship.

Moreover, it is obvious that state logics can differ to a considerable degree. They can be more open (e.g. in populist regimes) or closed (e.g. in autocratic regimes) to contestation and develop other 'condensations' than the ones I have described. As a consequence, very different commodification gaps could be found.

Second, the case of super-gentrification demands more attention. My research on Barnsbury has already highlighted the use of gentrification as an asset strategy intermingled with the financialisation and globalisation of property markets. The use of local properties as global financial products, however, raises questions about a possible scalar disconnect between local regulatory landscapes targeted at the residential sector and global investment strategies. Limiting commodification is, moreover, not restricted to policies targeting the conditions for the consumption of housing, but can also apply to the conditions of its production, e.g. in the form of mortgage regulations, credit scoring, bank supervision, etc. Studying the commodification gap would, therefore, clearly profit from a more intensive engagement with the literature on the financialisation of real estate that has progressed greatly in the past decade (for a literature review, see Aalbers 2019).

Third, it would be interesting to apply the concept of the commodification gap to areas beyond housing. Urban living consists of more than accommodation and there have been excellent contributions on commercial gentrification (Gonzalez and Waley 2013), the gentrification of public places (Mitchell 2003; Modan 2007), food gentrification (Alkon and Cadji 2020; Hope et al. 2020), and other fields that are not easily covered by a concept tailored to housing. While all these follow different logics than the ones described here, I would also maintain that a nexus between commodification and decommodification can be found in them too.

Finally, putting the state into the centre of the analysis may open the door for gentrification research that is more interested in history and political determination (for excellent examples see van Gent and Boterman 2019; Hackworth and Smith 2001). Taking the political determination of gentrification seriously can serve as a reminder against over-simplification at a minimum, but it also provides opportunities for a wide range of new research, reaching from systematic comparison of different housing and welfare regimes with regard to their impact on sociospatial inequalities and segregation, to studies on political conflict structures and opportunities for contestation, to the reflection of gentrification in public discourse, electoral strategies and coalition building.

Closely related to this call for a more politically interested gentrification research, this book also suggests a different perspective on the implications of gentrification theory for progressive political strategies and activism. Whereas I agree with many radical scholars that in the long run the only defence against gentrification is the decommodification of housing, I have my reservations about the idea that this will only be possible when 'the needs of capital would be systematically dismantled, to be displaced by the social, economic, and cultural needs of people' (Smith 1979, p. 547) or that change 'is unlikely to be achieved through a series of reforms'[2] (Lees et al. 2008, p. 222). The experiences made in the three cases studied not only demonstrate how gentrification is proceeding and social rights are being subordinated to the needs of capital, but they have also pointed towards actual experiences with partial decommodification on a systemic level. Real-world alternatives to gentrification are already there, and they are embedded in the histories of housing in the three societies discussed. There is no silver bullet for unmaking gentrification and change will not be achieved through a one-size-fits-all approach. What is needed then, is not abstract anti-capitalism and Manichean utopias, but a better understanding of the political forces that can make alternatives to gentrification possible. Political institutions are man-made. They are the outcome of political struggles. Better understanding the ways in which these are connected to gentrification can support diagnosis, enable better strategies, and guide imagining and organising for more equitable and just cities. Change will not come in the form of one single regulation, planning tool or policy and there will never be a single, universally valid, theoretical device on the basis of which sound strategies will be developed. Actually, intervening ideas can hardly be developed as a ready-made blueprint written by academics and 'given' to real-world actors. Instead, they should be developed in close relation to given political constellations and stimulate action. The commodification gap developed in this book is meant to provide some guidance for how this can be done in a practical way that at the same time is theoretically informed.

In this context, developing alternatives can be supported by systematically expanding one's horizon. Focusing on the institutional, political and cultural underpinnings of gentrification is, therefore, far more than 'diversions in an epoch of vicious state-led accumulation strategies, and the ever-sophisticated mutation of neoliberal urbanism' (Slater 2018, p. 124). If gentrification is made through the closing of commodification gaps, it can also be unmade by the reverse process. Understanding the commodification/decommodification nexus is not a revisionist deviation, but a necessary precondition for paving a path from the abstract world of political economy into the actually existing jungle of political struggles. At a time when housing provision is in crisis in many Western and non-Western cities and gentrification seems to be ubiquitous, comparing experiences of different forms of decommodification can help us gain a better sense of what is possible. Why should English private tenants not look at German rent laws? Or, why should Germans not allow themselves to be informed by the experience with the Right to Buy Policy when it comes to evaluating whether individual homeownership can work as a means to prevent gentrification? In short, cross-context learning can help us spot real-world alternatives, and also assist in uncovering false alternatives.

It was the goal of this study to reaffirm the primacy of the political in gentrification. As such, political choices are always in flux, and so is gentrification. Real solutions towards gentrification will only be found in real-world experiments. The conclusions of this study must, therefore, remain open-ended. The key to overcoming gentrification is political. Gentrification, like most societal phenomena, is a matter of 'Who gets what, when, how' (Lasswell 1936). It is my hope that this study will be regarded as suggesting one way in which fruitful thinking about this question can be made possible.

Notes

1. Closer examination, or the inclusion of other cases in other institutional environments, might help us identify even more gaps.

2. Interestingly, in more recent publications, Loretta Lees suggests a series of 'reformist' steps (see Lees et al. 2016, pp. 222ff.). The position cited above, thus, shouldn't be seen as dogmatic.

Appendix A

Compulsory Purchase in Barnsbury

TABLE A.1 Housing units acquired through the use of compulsory purchase orders by the borough of Islington between 1973 and 1976 and sold after 1995 in the Barnsbury Ward (data retrieved from Council of Islington 1973–1976 and UK Land Registry Open Data, http://landregistry.data.gov.uk/, retrieved 5 December 2016)

Address	Date of sale	Price paid (£)	Property type
190 Liverpool Road	6 Dec 2013	1 200 000	New built terrace
180 Liverpool Road	3 Sep 2013	1 250 000	New built terrace
182 Liverpool Road	9 Aug 2013	1 210 000	New built terrace
184 Liverpool Road	19 Jul 2013	1 420 000	New built terrace
172 Liverpool Road, Flat 1	3 Jun 2013	340 000	New built flats
172 Liverpool Road, Flat 2	3 Jun 2013	375,000	New built flats
172 Liverpool Road, Flat 3	17 Jun 2013	540 000	New built flats
172 Liverpool Road, Flat 4	19 Jul 2013	323 000	New built flats
172 Liverpool Road, Flat 5	3 Jun 2013	485 500	New built flats
172 Liverpool Road, Flat 6	4 Jun 2013	390 000	New built flats
172 Liverpool Road, Flat 7	3 Jun 2013	390 000	New built flats
172 Liverpool Road, Flat 8	3 Jun 2013	360 000	New built flats
172 Liverpool Road, Flat 9	24 Jun 2013	508 000	New built flats
	17 Dec 2013	675 000	

(Continued)

The Commodification Gap: Gentrification and Public Policy in London, Berlin and St Petersburg, First Edition. Matthias Bernt.
© 2022 John Wiley & Sons Ltd. Published 2022 by John Wiley & Sons Ltd.

TABLE A.1 (Continued)

Address	Date of sale	Price paid (£)	Property type
172 Liverpool Road, Flat 10	10 Jun 2013	470 000	New built flats
172 Liverpool Road, Flat 11	10 Jun 2013	800 000	New built flats
172 Liverpool Road, Flat 12	3 Jun 2013	915 000	New built flats
174 Liverpool Road	20 Dec 2013	1 350 000	New built terrace
176 Liverpool Road	10 Mar 2014	1 450 000	New built terrace
6 Copenhagen Street	21 Mar 1997	129 500	Terrace
6a Copenhagen Street	17 Jul 1997	170 000	Flat maisonette
6b Copenhagen Street	30 Mar 1995	97 000	Flat maisonette
	5 May 1998	181 000	
10a Copenhagen Street	31 May 1996	108 500	Flat maisonette
10b Copenhagen Street	6 Dec 1995	118 000	Flat maisonette
	25 Sep 1998	215 000	
17c, Copenhagen Street	7 Jul 2015	462 595	Flat maisonette
21Copenhagen Street	8 Aug 1995	150 000	Terrace
	23 May 1997	335 000	
	24 Apr 2006	657 000	
27 Lonsdale Square	8 Jan 1999	205 000	Flat maisonettes
Flat	28 Jan 2005	420 000	
22a Lonsdale Square	21 Oct 2015	1 435 000	Terraced
22b Lonsdale Square	20 Jul 2005	293 000	Flat maisonette
Albert Mansions, 357 Liverpool Road, Flat 4	2 Oct 2009	205 250	Flat maisonettes
Albert Mansions, 357 Liverpool Road, Flat 7	11 Apr 2012	283 000	Flat maisonettes
Albert Mansions, 357 Liverpool Road, Flat 2	30 Oct 2013	345 000	Flat maisonettes

TABLE A.1 (Continued)

Address	Date of sale	Price paid (£)	Property type
Avon House, Offord Road, Flat 3	25 Jul 2011	312 500	Flat maisonettes
	13 May 2016	595 000	
Avon House, Offord Road, Flat 6	5 May 1995	57 500	Flat maisonettes
	26 March 1999	120 000	
	10 November 2000	170 000	
		338 000	
	16 November 2012	475 000	
	20 January 2014	590 000	
	7 Aug 2014		
Avon House, Offord Road, Flat 8	19 Jul 2002	191 000	Flat maisonettes
Avon House, Offord Road, Flat 10	29 March 2010	308 000	Flat maisonettes
Bures House, Offord Road, Flat 5	21 August 2001	135 000	Flat maisonettes
	1 November 2004	245 000	
		379 000	
	17 October 2011		
Bures House, Offord Road, Flat 6	25 April 2002	60 000	Flat maisonettes
Bures House, Offord Road, Flat 7	26 August 2005	241 000	Flat maisonettes
	1 August 2008	359 000	
Buckland House, Offord Road, Flat 2	9 May 2003	250 000	Flat maisonettes
	30 April 2012	370 000	
Buckland House, Offord Road, Flat 4	23 Jun 1999	130 000	Flat maisonettes
	10 Aug 2005	235 560	
Buckland House, Offord Road, Flat 5	3 Sept 2008	247 500	Flat maisonettes
Buckland House, Offord Road, Flat 7	18 Jun 2010	295 000	Flat maisonettes
Thornhill House, Thornhill Road, Flat 53	28 Apr 1995	107 000	Flat maisonettes

(Continued)

TABLE A.1 (Continued)

Address	Date of sale	Price paid (£)	Property type
Thornhill House, Thornhill Road, Flat 59	13 Jan 1997	68 500	Flat maisonettes
Thornhill House, Thornhill Road, Flat 62	5 Mar 1999	90 000	Flat maisonettes
	29 Sep 2000	100 000	
	20 Oct 2004	210 000	
Thornhill House, Thornhill Road, Flat 64	17 Jan 2011	219 000	Flat maisonettes
Thornhill House, Thornhill Road, Flat 65	28 Aug 1996	64 950	Flat maisonettes
	17 Mar 2006	220 000	
	11 Dec 2009	260 000	
Thornhill House, Thornhill Road, Flat 68	6 Jun 2006	219 000	Flat maisonettes
Thornhill House, Thornhill Road, Flat 69	1 Sep 2003	182 500	Flat maisonettes
Thornhill House, Thornhill Road, Flat 71	21 Sep 2007	250 000	Flat maisonettes
Thornhill House, Thornhill Road, Flat 74	31 Oct 1995	83 000	Flat maisonettes
	7 Dec 2001	207 000	
Thornhill House, Thornhill Road, Flat 82	1 Sep 2000	135 000	Flat maisonettes
	14 Jul 2006	214 300	
	6 Nov 2015	455 000	
18, Barnsbury Road, Flat 2	18 Jul 1995	68 000	Flat maisonettes
18, Barnsbury Road, Flat 3	25 Nov 2014	400 000	Flat maisonette
18, Barnsbury Road, Flat 5	12 Nov 2007	319 995	Flat maisonette
	1 Mar 2013	325 000	

TABLE A.1 (Continued)

Address	Date of sale	Price paid (£)	Property type
163 Barnsbury Road	19 Mar 2002	526 000	Terrace
	9 Apr 2015	1 500 000	
165 Barnsbury Road	16 Jul 2015	1 490 000	Terrace
167 Barnsbury Road	19 May 2005	610 000	Terrace
	19 June 2006	650 000	
169 Barnsbury Road	28 June 2005	782 500	Terrace
171 Barnsbury Road	19 May 2005	650 000	Terrace
183b Barnsbury Road	28 Mar 1996	104 500	Flat maisonettes
	13 Nov 1997	193 000	
	20 Sep 2002	280 000	
	29 Sep 2005	325 000	
183c Barnsbury Road	13 Nov 2014	855 000	Flat maisonettes
173 Barnsbury Road	4 Feb 1999	360 000	Terrace
Terrace	20 Oct 2000	573 000	
	31 May 2002	662 500	
	22 Sep 2005	765 000	
89a Richmond Avenue	23 Jan 2004	525 000	Flat maisonettes
90a Richmond Avenue	4 Mar 2015	262 000	Flat maisonette
61a Thornhill Square	23 Sep 1998	165 000	Flat maisonettes
	17 Mar 2000	232 500	
124 Cloudesley Road	29 May 2009	1 035 000	Terraced

Appendix B
Residents in NS-SeC Classes 1 and 2

The Commodification Gap: Gentrification and Public Policy in London, Berlin and St Petersburg, First Edition. Matthias Bernt.
© 2022 John Wiley & Sons Ltd. Published 2022 by John Wiley & Sons Ltd.

TABLE B.1 Share of NS-SeC Class 1 and 2 and tenure as a percent of all residents aged 16–74 in UK Census Output Areas of Barnsbury in 2011 (UK Census 2011, http://www.ukcensusdata.com/barnsbury-e05000366)

UK Census Output Area (from south to north)	NS-SeC/Class 1 in percent of all residents aged 16-74	NS-SeC/Class 2 in percent of all residents aged 16-74		Tenure in percent of all households (approximate values)							
				Owned: Owned outright	Owned: Owned with a mortgage or loan	Shared ownership (part owned and part rented)	Social rented: Rented from council (local authority)	Social rented: Other	Private rented: Private landlord or letting agency	Private rented: Other	Living rent free
Barnsbury E00013385	14.6	21.4	36.0	3.2	11.1	0.0	6.9	50.3	27.5	0.5	0.5
Barnsbury E00013386	28.6	32.9	61.5	27.6	26.7	1.7	23.3	0.9	17.2	0.9	1.7
Barnsbury E00013387	16.5	27.1	43.6	6.0	2.6	0.0	0.86	70.1	17.1	3.4	0.0
Barnsbury E00013388	23.1	26.9	50.0	10.4	20.0	0.0	43.7	8.9	15.6	1.5	0.0
Barnsbury E00013389	38	30.7	68.7	23.0	25.0	1.0	7.0	10.0	28.0	2.0	4.0
Barnsbury E00013390	35.1	28.5	63.6	10.3	14.7	9.8	1.6	15.8	43.5	3.3	1.1
Barnsbury E00013391	7.4	25.0	32.4	7.3	5.8	0.8	12.4	59.9	10.9	0.7	2.2

Barnsbury E00013392	6.0	11.0	17.0	5.0	3.4	0.0	11.8	63.0	13.4	1.7	1.7
Barnsbury E00013393	4.3	13.0	17.3	2.0	4.0	0.0	8.0	72.0	12.0	0.0	2.0
Barnsbury E00013395	15.7	23.1	38.8	6.9	6.9	0.0	7.9	63.4	10.9	1.0	3.0
Barnsbury E00013397	25.9	34.2	60.1	12.3	24.3	0.0	10.1	12.8	37.0	2.2	0.9
Barnsbury E00013398	28.3	31.4	59.7	17.2	31.1	0.0	18.0	6.6	25.4	1.6	0.0
Barnsbury E00013399	24.1	28.8	52.9	9.2	24.4	0.0	26.7	19.4	19.4	0.4	0.4
Barnsbury E00013400	40.0	24.6	64.6	15.7	15.7	0.0	41.7	2.8	22.2	0.9	0.9
Barnsbury E00013401	30.6	37.0	67.6	25.8	30.0	0.0	7.5	5.0	29.2	0.8	1.7
Barnsbury E00013402	7.5	21.7	29.2	3.0	3.0	0.0	4.1	88.8	1.0	0.0	0.0
Barnsbury E00013403	35.7	27.6	63.3	15.3	24.3	0.0	13.9	10.9	34.3	0.7	0.0
Barnsbury E00013404	35.8	33.2	69.0	34.6	31.5	0.0	3.1	4.7	24.6	0.0	2.4
Barnsbury E00013405	24.8	43.1	67.9	16.8	24.8	0.7	15.3	5.8	34.3	1.4	0.7

(Continued)

TABLE B.1 (Continued)

UK Census Output Area (from south to north)	NS-SeC/Class 1 in percent of all residents aged 16–74	NS-SeC/Class 2 in percent of all residents aged 16–74		Tenure in percent of all households (approximate values)							
				Owned: Owned outright	Owned: Owned with a mortgage or loan	Shared ownership (part owned and part rented)	Social rented: Rented from council (local authority)	Social rented: Other	Private rented: Private landlord or letting agency	Private rented: Other	Living rent free
Barnsbury E00013406	38.0	33.1	71.1	28.2	26.5	0.0	5.3	3.5	35.3	1.2	0.0
Barnsbury E00013407	40.4	33.0	73.4	52.3	22.0	0.0	3.7	2.7	17.4	0.9	0.9
Barnsbury E00013408	40.9	28.3	69.2	35.9	28.8	0.0	7.8	2.0	21.6	0.6	3.3
Barnsbury E00013409	28.9	26.1	55.0	15.6	23.1	0.0	34.7	6.6	19.1	0.6	0.6
Barnsbury E00013410	26.5	34.5	61.0	34.5	28.3	0.0	6.2	3.5	25.7	1.8	0.0
Barnsbury E00013411	19.6	21.7	41.3	6.4	12.8	0.0	7.9	47.9	21.4	1.4	2.1
Barnsbury E00013412	25.4	25.7	51.1	18.8	22.5	0.0	25.3	2.9	27.5	0.7	2.2

Barnsbury E00013413	39.6	32.6	72.2	34.3	24.3	0.0	1.8	10.8	27.9	0.9	0.0
Barnsbury E00013414	14	17.6	31.6	1.3	7.3	11.3	11.9	55.0	10.6	0.6	2.0
Barnsbury E00013415	6.5	16.9	23.4	5.3	5.3	0.0	8.8	61.9	12.4	3.5	2.7
Barnsbury E00013416	8.4	16.7	25.1	2.9	6.6	0.0	10.3	65.5	11.0	0.7	2.9
Barnsbury E00013417	7.6	29.0	36.6	10.2	8.8	1.5	5.1	52.6	18.2	2.2	1.5
Barnsbury E00013418	6.9	14.9	21.8	2.3	7.0	0.0	6.2	72.9	11.6	0.0	0.0
Barnsbury E00013419	26.1	33.6	59.7	9.7	13.5	11.4	3.8	5.9	53.5	1.1	1.1
Barnsbury E00013420	34.1	27.6	61.7	16.2	22.5	1.0	13.6	12.6	32.4	1.0	0.5
Barnsbury E00013421	11	20.3	31.3	9.8	15.5	0.7	14.8	43.7	13.4	0.7	1.4
Barnsbury E00013422	28.7	35.9	64.6	4.6	7.7	0.0	4.6	1.3	75.5	4.0	2.0
Barnsbury E00013423	24.7	21.6	46.3	10.1	19.4	1.6	16.2	23.3	27.9	0.0	1.5
Barnsbury E00174805	16	22.3	38.3	7.4	8.3	0.0	4.4	51.5	23.5	1.5	2.9

(Continued)

TABLE B.1 (Continued)

UK Census Output Area (from south to north)	NS-SeC/Class 1 in percent of all residents aged 16–74	NS-SeC/Class 2 in percent of all residents aged 16–74		Tenure in percent of all households (approximate values)							
				Owned: Owned outright	Owned: Owned with a mortgage or loan	Shared ownership (part owned and part rented)	Social rented: Rented from council (local authority)	Social rented: Other	Private rented: Private landlord or letting agency	Private rented: Other	Living rent free
Barnsbury E00174836	32.6	32.2	64.8	15.9	28.0	0.5	11.5	4.9	35.2	2.2	1.6
Barnsbury E00174843	4.3	5.5	9.8	6.4	4.2	0.0	4.2	10.7	72.3	2.1	0.0
Barnsbury E00174847	22.9	25.9	48.8	6.3	10.0	0.0	16.3	37.5	28.8	1.2	0.0
Barnsbury	–	–	–	14.2	17.5	1.2	12.3	26.2	26.0	1.3	1.2
London	13.2	23.1	36.3	21.1	27.1	1.3	13.4	10.6	23.7	1.3	1.3

References

Aalbers, M.B. (2012). *Subprime Cities: The Political Economy of Mortgage Markets*. Oxford: Wiley Blackwell.

Aalbers, M.B. (2016). *The Financialization of Housing: A Political Economy Approach*. London: Routledge.

Aalbers, M.B. (2019). Financial Geographies of Real Estate and the City. A Literature Review (Financial Geography Working Paper #21). http://www.fingeo.net/wordpress/wp-content/uploads/2019/01/FinGeoWP_Aalbers-2019-2.pdf (accessed 23 March 2021).

AHML. Agency for Housing and Mortgage Lending Research Centre (2015). Housing and the Mortgage Lending Market. Results for 2014. http://www.ahml.ru/common/img/uploaded/files/agency/reporting/quarterly/report4q2014_en.pdf (accessed 21 June 2016; webpage no longer available).

Alkon, A.H. and Cadji, J. (2020). Sowing seeds of displacement: Gentrification and food justice in Oakland, CA. *International Journal of Urban and Regional Research* 44: 108–123.

Allen, C. (2008). *Housing Market Renewal and Social Class*. London: Routledge.

Andrusz, G.D. (1987). The built environment in Soviet theory and practice. *International Journal of Urban and Regional Research* 11(4): 478–499.

Arbaci, S. (2019). *Paradoxes of Segregation: Housing Systems, Welfare Regimes and Ethnic Residential Change in Southern European Cities*. Oxford: Wiley.

ASUM. Angewandte Sozialforschung und Urbanes Management GmbH (2012). Endbericht zur Sozialstudie Sanierungsgebiet Prenzlauer Berg – Helmholtzplatz 2012 [Final Report on the Study of Social Structures in the Urban Renewal Area Prenzlauer Berg – Hemholtzplatz 2012]. https://www.berlin.de/imperia/md/content/bapankow/stapl/sozialstudie_helmholtzplatz_2012.pdf (accessed 10 June 2016; webpage no longer available).

Atkinson, R., Parker, S., and Burrows, R. (2017). Elite formation, power and space in contemporary London. *Theory, Culture & Society* 34(5–6): 179–200.

Axënov, K. (2011). Social segregation of personal activity spaces in a post-transformation metropolis (case study of St. Petersburg). *Regional Research of Russia* 1(1): 52–61.

Badcock, B. (1989). An Australian view of the rent gap hypothesis. *Annals of the Association of American Geographers* 79(1): 125–145.

Badcock, B. (1990). On the nonexistence of the rent gap, a reply. *Annals of the Association of American Geographers* 80(3): 459–461.

Badyina, A. and Golubchikov, O. (2005). Gentrification in central Moscow – A market process or deliberate policy? Money, power and people in housing regeneration in Ostozhenka. *Geografiska Annaler: Series B Human Geography* 87(2): 113–129.

Balchin, P. (1996). *Housing Policy in Europe*. London: Routledge.

Ball, M. (1977). Differential rent and the role of landed property. *International Journal of Urban and Regional Research* 1(1–4): 380–403.

Ball, M. (1985a) Land rent and the construction industry. In: *Land Rent: Housing and Urban Planning* (eds. M. Ball, V. Bentivegna, and M. Edwards), 71–86. London: Groom Helm.

Ball, M. (1985b). The urban rent question. *Environment and Planning A* 17: 503–525.

Ball, M. (1987). Rent and social relations: A reply to Clark. *Environment and Planning A* 19: 269–272.

Ball, M. (1998). Institutions in British property research: A review. *Urban Studies* 35(9): 1501–1517.

Barlow, J. and Duncan, S. (1994). *Success and Failure in Housing Provision: European Systems Compared*. Oxford: Pergamon.

Bartholomäi, R. (2004). Die Entwicklung des Politikfeldes Wohnen [The development of housing policy]. In: *Housing Policy in Germany. Positions, Agents, Instruments* (eds. B. Egner, H. Georgakis, H. Heinelt, and R. Bartholomäi), 15–34. Kassel: Schader Stiftung.

Bater, J.H. (1980). *The Soviet City: Ideal and Reality*. Beverly Hills, CA: Sage.

Bater, J.H. (2006). Central St. Petersburg: Continuity and change in privilege and place. *Eurasian Geography and Economics* 47(1): 4–27.

Beauregard, R.A. (1986). The chaos and complexity of gentrification. In: *Gentrification of the City* (eds. N. Smith and P. Williams), 35–55. Winchester: Allen and Unwin.

Beauregard, R.A. (2003). *City of superlatives. City and Community* 2(3): 183–199.

Becker, R. (1988). Subventionen für den Wohnungssektor [Subsidies for the housing sector]. In: *Sozialer Wohnungsbau im internationalen Vergleich [Social Housing in International Comparison]* (eds. W. Prigge and W. Kaib), 94–122. Frankfurt a.M.: Vervuert Verlag.

Bengtsson, B. (2015). Between structure and Thatcher: Towards a research agenda for theory-informed actor-related analysis of housing politics. *Housing Studies* 30(5): 677–693.

Bernt, M. (1998). *Stadterneuerung unter Aufwertungsdruck [Urban Renewal and Upgrading Pressure]*. Bad Sinzheim: Pro-Universitate-Verlag.

Bernt, M. (2003). *Rübergeklappt: Die 'Behutsame Stadterneuerung' im Berlin der 1990er Jahre [Flipped. Careful Urban Renewal in Berlin in the 1990s]*. Berlin: Schelzky & Jeep.

Bernt, M. (2009). Partnerships for demolition: The governance of urban renewal in East Germany's shrinking cities. *International Journal of Urban and Regional Research* 33(4): 754–769.

Bernt, M. (2012). The 'double movements' of neighbourhood change: Gentrification and public policy in Harlem and Prenzlauer Berg. *Urban Studies* 49(14): 3045–3062.

Bernt, M. (2015). Stadterneuerung zwischen Entstaatlichung und politischer Einflußnahme. Zu schwach, zu spät, zu unentschlossen [Urban renewal between state-withdrawal and political interventionism. Too late, too weak, too indecisive]. In: *Eine Stadt verändert sich. Berlin-Pankow 25 Jahre Stadterneuerung [A city changes. Berlin-Pankow 25 years of Urban Renewal]* (ed. Bezirksamt Pankow. Abteilung Stadtentwicklung), 150–161. Berlin: Nicolai.

Bernt, M. (2016a). Very particular, or rather universal? Gentrification through the lenses of Ghertner and López-Morales. *City* 20(4): 637–644.

Bernt, M. (2016b). How post-socialist is gentrification? Observations in East Berlin and Saint Petersburg. *Eurasian Geography and Economics* 57(4–5): 565–587.

Bernt, M. (2017). Phased out, demolished and privatized: Social housing in an East German shrinking city. In: *Social Housing and Urban Renewal: A Cross-National Perspective* (eds. P. Watt and P. Smets), 253–276. Oxford: Emerald Publishing Ltd.

Bernt, M. and Holm, A. (2005). Exploring the substance and style of gentrification: Berlin's 'Prenzlberg.' In: *Gentrification in a Global Perspective: The New Urban Colonialism* (eds. R. Atkinson and G. Bridge), 106–120. Milton Park: Routledge.

Bernt, M. and Holm, A. (2009). Is it, or is not? The conceptualisation of gentrification and displacement and its political implications in the case of Berlin Prenzlauer Berg. *City* 13(2–3). 312–324.

Bernt, M., Colini, L., and Förste, D. (2017). Privatization, financialization and state restructuring in East Germany: The case of Am Südpark. *International Journal of Urban and Regional Research* 41(4): 555–571.

Betancur, J. (2014). Gentrification in Latin America: Overview and critical analysis. *Urban Studies Research*. http://dx.doi.org/10.1155/2014/986961.

Bezirksamt Pankow. Abteilung Stadtentwicklung (2015). *Eine Stadt verändert sich. Berlin-Pankow 25 Jahre Stadterneuerung [A city changes. Berlin-Pankow 25 years of Urban Renewal]*. Berlin: Nicolai.

Bhaskar, R. (1975). *A Realist Theory of Science*. Leeds: Leeds Books.

Binney, M. (2012). Introduction. In: *Sankt Peterburg: nasledye pod ugrozoy [St Petersburg: Heritage at Risk]* (eds. C. Cecil and E. Minchenok), 13–15. Moscow: MAPS.

Blomley, N. (2003). Law, property, and the geography of violence: The frontier, the survey, and the grid. *Annals of the Association of American Geographers* 93(1): 121–141.

Blomley, N. (2004). *Unsettling the City: Urban Land and the Politics of Property*. New York, NY: Routledge.

Blomley, N. (2005). Flowers in the bathtub: Boundary crossings at the public–private divide. *Geoforum* 36(3): 281–296.

Bodenschatz, H. (1987). *Platz frei für das neue Berlin! Geschichte der Stadterneuerung in der 'größten Mietskasernenstadt der Welt' seit 1871 [Give Way to the New Berlin! A History of Urban Renewal in the 'Biggest Tenement City of the World']*. Berlin: Transit.

Bondi, L. (1999). Between the woof and the weft: A response to Loretta Lees. *Environment and Planning D: Society and Space* 17(3): 125–145.

Bonefeld, W. (2012). Freedom and the strong state: On German ordoliberalism. *New Political Economy* 17(5): 633–656.

Borst, R. (1996): Volkswohnungsbestand in Spekulantenhand? [People's homes for speculators?]. In: *Stadtentwicklung in Ostdeutschland [Urban development in East Germany]*(eds. H. Häußermann and R. Neef), 107–128. Opladen: Westdeutscher Verlag.

Bourassa, S.C. (1993). The rent gap debunked. *Urban Studies* 30(10): 1731–1744.

Bowie, D. (2016). *The Radical and Socialist Tradition in British Planning: From Puritan Colonies to Garden Cities*. London: Routledge.

Bowie, D. (2017). Beyond the compact city: A London case study – Spatial impacts, social polarisation, sustainable development and social justice (International Eco-Cities Initiative 1), https://www.researchgate.net/publication/324080429_Beyond_the_compact_city_a_London_case_study_-_spatial_impacts_social_polarisation_sustainable_development_and_social_justice (accessed 23 October 2020).

Brenner, N. (2003). Stereotypes, archetypes and prototypes: Three uses of superlatives in contemporary urban studies. *City and Community* 2(3): 205–216.

Brenner, N. and Theodore, N. (2002). Cities and the geographies of 'actually existing neoliberalism'. *Antipode* 34(3): 349–379.

Bridge, G. and Dowling, R. (2001). Microgeographies of retailing and gentrification. *Australian Geographer* 32(1): 93–107.

Bridge, G., Butler, T., and Lees, L. (2012). *Mixed Communities. Gentrification by Stealth?* Bristol: Policy Press.

Brown-Saracino, J. (2010). *The Gentrification Debates*. New York, NY: Routledge.

Bugler, J. (1968). The invaders of Islington. *New Society*, 15 August.

Butler, S.B., Nayyar-Stone, R., and O'Leary, S. (1999). The law and economics of historic preservation in St. Petersburg, Russia. *Review of Urban & Regional Development Studies* 11(1): 24–44.

Butler, T. (2007). For gentrification? *Environment and Planning A* 39(1): 162–181.

Butler, T. and Lees, L. (2006). Super-gentrification in Barnsbury, London: Globalization and gentrifying global elites at the neighbourhood level. *Transactions of the Institute of British Geographers* 31: 467–487.

Butler, T. and Robson, G. (2003). *London Calling. The Middle Classes and the Re-Making of Inner London*. Oxford: Berg.

Calbet I Elias, L. (2018). Financialised rent gaps and the public interest in Berlin's housing crisis. Reflections on N. Smith's 'generalised gentrification'. In:

Gentrification as a Global Strategy. Neil Smith and Beyond (eds. A. Albet and N. Benach), 165–176. London and New York, NY: Routledge.

Callon, M. (1998). *The Laws of the Markets*. London: Blackwell Publishers.

Castles, F.G. (1998). *Comparative Public Policy. Patterns Post-War Transformation*. Cheltenham: Edward Elgar.

Castles, F.G. (2005). The Kemeny thesis revisited. *Housing, Theory and Society* 22(2): 84–86.

Chakrabarty, D. (2000) *Provincializing Europe: Postcolonial Thought and Historical Difference*. Princeton, NJ: Princeton University Press.

Chambers, P.A.B. (1974) The process of gentrification in Inner London. M.Phil. thesis. School of Environmental Studies, University College of London.

Choi, N. (2016). Metro Manila through the gentrification lens: Disparities in urban planning and displacement risks. *Urban Studies* 53(3) (Special issue: Locating gentrification in East Asia): 577–592.

Clark, E. (1987). *The Rent Gap and Urban Change: Case Studies in Malmö, 1860–1985*. Lund: Lund University Press.

Clark, E. (1988). The rent gap and the transformation of the built environment: Case studies in Malmö 1860–1985. *Geografiska Annaler B* 70: 241–254.

Clark, E. (1992). On gaps in gentrification theory. *Housing Studies* 7(1): 16–26.

Clark, E. (2005). The order and simplicity of gentrification – A political challenge. In: *Gentrification in a Global Context. The New Urban Colonialism* (eds. R. Atkinson and G. Bridge), 256–264. London: Routledge.

Clarke, S. and Ginsburg, N. (n.d.) The Political Economy of Housing. https://homepages.warwick.ac.uk/~syrbe/pubs/ClarkeGinsburg.pdf (accessed 1 April 2017).

Colomb, C. (2007). Unpacking New Labour's 'urban renaissance' agenda. Towards a socially sustainable reurbanization of British cities? *Planning Practice and Research* 22(1): 1–24.

Cook, L.J. (2007). *Postcommunist Welfare States. Reform Politics in Russia and Eastern Europe*. Ithaca, NY: Cornell University Press.

Copley, T. (2014). From Right to Buy to Buy to Let. Ed. Greater London Authority. http://tomcopley.com/wp-content/uploads/2014/01/From-Right-to-Buy-to-Buy-to-Let-Jan-2014.pdf (accessed 15 February 2015; webpage no longer available).

Corbett, S. and Walker, A. (2012). The big society: Back to the future. *The Political Quarterly* 83(3): 487–493.

Crook, T. and Kemp, P.A. (2011). *Transforming Private Landlords. Housing, Market and Public Policy*. Oxford: Wiley Blackwell.

Crouch, C. and Streeck, W. (1997). *Political Economy of Modern Capitalism*. London: Sage.

Cummings, J. (2016). Confronting favela chic. The gentrification of informal settlements in Rio de Janeiro, Brazil. In: *Global Gentrifications* (eds. L. Lees, H.B. Shin, and E. López-Morales), 81–100. Bristol and Chicago, IL: Policy Press.

Czada, R. (1985). Neokorporatistische Politikentwicklung in Westeuropa. Politische Verbändeeinbindung und wirtschaftspolitische Strategien im internationalen

Vergleich [Neocorporatist Policy Development in Western Europe], Research report given to Stiftung Volkswagenwerk. http://www.politik.uni-osnabrueck.de/POLSYS/publications.html (accessed 20 November 2017).

Czada, R. (2000). Konkordanz, Korporatismus und Politikverflechtung: Dimensionen der Verhandlungsdemokratie [Concordance, corporatism and joint decision-making: Dimensions of the negotiating democracy]. In: *Zwischen Wettbewerbs- und Verhandlungsdemokratie. Analysen zum Regierungssystem der Bundesrepublik Deutschland [Between Competition and Negotiations. Analyses on the Governance Regime of the Federal Republic of Germany]* (eds. E. Holtmann and H. Voelzkow), 23–52. Wiesbaden: Westdeutscher Verlag.

Czada, R. (2003). Der Begriff der Verhandlungsdemokratie und die vergleichende Policy-Forschung [The concept of a negotiating democracy in comparative policy research]. In: *Die Reformierbarkeit der Demokratie. Innovationen und Blockaden [The Reformability of Democracy. Innovations and Blockades]* (eds. R. Mayntz and W. Streeck), 173–203. Frankfurt: Campus.

Czada, R. and Schmidt, M.G. (1993). *Verhandlungsdemokratie, Interessenvermittlung, Regierbarkeit [Negotiating Democracy, Mediation of Interests and Governability]*. Wiesbaden: Westdeutscher Verlag.

Daunton, M.J. (1987). *A Property-Owning Democracy? Housing in Britain*. London and Boston, MA: Faber and Faber.

Davidson, M. and Lees, L. (2005). New-build 'gentrification' and London's riverside renaissance. *Environment and Planning A* 37: 1160–1195.

Davidson, M. and Lees, L. (2010). New-build gentrification. Its histories, trajectories, and critical geographies. *Population, Space and Place* 16: 395–411.

Davis, J. (2001). Rents and race in 1960s London: New light on Rachmanism. *Twentieth Century British History* 12(1): 69–92.

DCH. Defend Council Housing. (2008). Briefing. The Case Against Transfer. http://www.defendcouncilhousing.org.uk/dch/resources/StockTransfer/DCHStockTransferbriefing.pdf (accessed 20 May 2014).

Delgadillo, V. (2015). Selective modernization of Mexico City and its historic center. Gentrification without displacement? *Urban Geography* 37(8) (Special Issue: Latin American gentrifications): 1154–1174.

DETR. Department of Environment, Transport and the Regions. (2000). *Quality and Choice: A Decent Home for All. The Housing Green Paper*. London: HSMO.

Diamond, L. (2002). Thinking about hybrid regimes. *Journal of Democracy* 13(2): 21–35.

Diappi, L. and Bolchi, P. (2008). Smith's rent gap theory and local real estate dynamics: A multi-agent model. *Computers, Environment and Urban Systems* 32: 6–18.

Dieser, H. (1996). Restitution – was ist sie und was bewirkt sie? [Restitution – what is it and what does it affect?]. In: *Stadtentwicklung in Ostdeutschland: Soziale und Räumliche Tendenzen [Urban development in East Germany. Social and spatial tendencies]* (eds. H. Häußermann and R. Neef), 107–128. Berlin: Leske & Budrich.

DoE. Department of the Environment. (1977). *Housing Policy. A Consultative Document*. London: HSMO.

Dörfler, T. (2010). *Gentrification in Prenzlauer Berg? Milieuwandel eines Berliner Sozialraums seit 1989 [Gentrification in Prenzlauer Berg? The Change of Social Milieux in a Neighbourhood in Berlin]*. Bielefeld: Transcript Verlag.

Doling, J. (1997). *Comparative Housing Policy. Government and Housing in Advanced Industrialized Countries*. London: Macmillan.

Donnison, D. (1967). *The Government of Housing*. Harmondsworth: Penguin.

Doshi, S. (2013) The politics of the evicted: Redevelopment, subjectivity, and difference in Mumbai's slum frontier. *Antipode* 45(4): 844–865.

Doshi, S. (2015). Rethinking gentrification in India: Displacement, dispossession and the spectre of development. In: *Gentrifications Global* (eds. L. Lees, H.B. Shin, and E. López-Morales), 101–120. Bristol and Chicago, IL: Policy Press.

Edwards, M. (2016). The housing crisis and London. *City* 20(2): 222–237.

Egner, B. (2014). Wohnungspolitik seit 1945 [Housing policy since 1945]. *Aus Politik und Zeitgeschichte* 20–21: 13–19.

Egner, B., Georgakis, H., Heinelt, H., and Bartholomäi, R. (2004). *Housing Policy in Germany. Positions, Agents, Instruments*. Kassel: Schader Stiftung.

Engels, B. (1994). Capital flows, redlining, and gentrification: The pattern of mortgage lending and social change in Glebe, Sydney, 1960–1984. *International Journal of Urban and Regional Research* 18(4): 628–657.

Esping-Andersen, G. (1990). *The Three Worlds of Welfare Capitalism*. Cambridge: Polity Press.

Etzioni, A. (1996). *The New Golden Rule: Community and Morality in a Democratic Society*. New York, NY: Basic Books.

Evans, P. (1995). *Embedded Autonomy. States and Industrial Transformation*. Princeton, NJ: Princeton University Press.

Fairclough, N. (2000). *New Labour, New Language?* London: Routledge.

Ferris, J. (1972). *Participation in Urban Planning: The Barnsbury Case*. Occasional Papers in Social Administration, No. 48. London: G. Bell and Sons.

Fligstein, N. (1990). *The Transformation of Corporate Control*. Cambridge, MA: Harvard University Press.

Fligstein, N. (2001). *The Architecture of Markets. An Economic Sociology of Twenty-First-Century Capitalist Societies*. Princeton, NJ: Princeton University Press.

Fligstein, N. and Calder, R. (2015). Architecture of markets. In: *Emerging Trends in the Social and Behavioral Sciences* (eds. R.A. Scott and S.M. Kosslyn). Hoboken, NJ: Wiley.

Fligstein, N. and Dioun, C., (2015). Economic sociology. In: *International Encyclopedia of the Social & Behavioral Sciences*, 2nd ed. Vol. 7 (ed.-in-chief J.D. Wright), 67–72. Oxford: Elsevier.

Forrest, R. and Murie, A. (1983). Residualization and council housing: Aspects of the changing social relations of housing tenure. *Journal of Social Policy*, 12(4): 453–468.

Forrest, R. and Murie, A. (1988). *Selling the Welfare State. The Privatisation of Public Housing*. London: Routledge.

French, R.A. (1995). *Plans, Pragmatism and People: The Legacy of Soviet Planning for Today's Cities*. Pittsburgh, PA: University of Pittsburgh Press.

French, R.A. and Hamilton, F.E.I. (1979). *The Socialist City: Spatial Structure and Urban Policy*. London: John Wiley.

Führer, K.C. (1995). *Mieter, Hausbesitzer, Staat und Wohnungsmarkt. Wohnungsmangel und Wohnungszwangswirtschaft in Deutschland 1914–1960 [Tenant, Landlord, State and Housing Market. Housing Shortage and Controlled Housing Economy in Germany 1914–1960]*. Stuttgart: Franz Steiner Verlag.

Furman, D. (2008). Imitation democracies: The post-Soviet penumbra. *New Left Review*, 54: 29–47.

Gentile, M., Salukvadze, J., and Gogishvili, D. (2015). Newbuild gentrification, tele-urbanization and urban growth: Placing the cities of the post-communist South in the gentrification debate. *Geografie* 120(2): 134–163.

Gerasimova, E.Y. (2000). *Sovetskaya kommunalnaya kvartira kak sotsialnyi institut: istoriko-sotsiologicheskii analiz material Petrograda-Leningrada 1917–1991* [The Soviet Communal Apartment as a Social Institution: Historical-Sociological Analysis of Material from Petrograd-Leningrad 1917–1991]. http://www.dissercat.com/content/sovetskaya-kommunalnaya-kvartira-kak-sotsialnyi-institut-istoriko-sotsiologicheskii-analiz-n (accessed 5 May 2016).

Gewos (2016). Bericht der Expertengruppe zur Reform des Sozialen Wohnungsbaus in Berlin [Report of the Experts' Commission on the Reform of Social Housing in Berlin]. http://www.stadtentwicklung.berlin.de/wohnen/sozialer_wohnungsbau/reform/download/teil01.pdf (accessed 20 March 2017).

Ghertner, D.A. (2014). India's urban revolution: Geographies of displacement beyond gentrification. *Environment and Planning A* 46(7): 1554–1571.

Ghertner, D.A. (2015). Why gentrification theory fails in 'much of the world'. *City* 19(4): 546–556.

Giddens, A. (1998). *The Third Way: The Renewal of Social Democracy*. Cambridge: Polity Press.

Giddens, A. (2000). *The Third Way and its Critics*. Cambridge: Polity Press.

Gillespie, T. (2020). The real estate frontier. *International Journal of Urban and Regional Research* 44(4): 599–616.

Ginsburg, N. (1989). The Housing Act, 1988 and its policy context: A critical commentary. *Critical Social Policy* 9(25): 56–81.

Ginsburg, N. (1999). Housing and housing policy in the 20th century. In: *British Social Welfare in the Twentieth Century* (eds. R. Page and R. Silburn), 226–235. UK: Palgrave Macmillan.

Ginsburg, N. (2005). The privatization of council housing. *Critical Social Policy* 25(1): 115–135.

Glass, R. (1964). *London: Aspects of Change*. London: McGibbon and Kee.

Glock, B. and Häußermann, H. (2004). New trends in urban development and public policy in eastern Germany: Dealing with the vacant housing problem at the local level. *International Journal of Urban and Regional Research* 28(4): 919–929.

Glucksberg, L. (2016). A view from the top. *City* 20(2): 238–255.

Glucksberg, L., Atkinson, R., Burrows, R. et al. (2016). Minimum city? A critical assessment of some of the deeper impacts of the 'super-rich' on urban life. In: *Cities and the Super-Rich: Real Estate, Elite Practices, and Urban Political Economies* (eds. R. Forrest, S. Yee Koh, and B. Wissink), 253–272. New York, NY: Palgrave Macmillan.

Goldschmidt, N. (2013). Walter Eucken's place in the history of ideas. *The Review of Austrian Economics* 26: 127–147.

Golubchikov, O. (2004). Urban planning in Russia: Towards the market. *European Planning Studies* 12(2): 229–247.

Gonzalez, S. and Waley, P. (2013). Traditional retail markets: The new gentrification frontier? *Antipode* 45(4): 965–983.

Goodchild, R. and Munton, R. (1985). *Development and the Landowner*. London: Allen & Unwin.

Gotham, K.F. (2005). Tourism gentrification: The case of New Orleans' Vieux Carre (French Quarter). *Urban Studies* 42(7): 1099–1121.

Gotham, K.F. (2009). Creating liquidity out of spatial fixity: The secondary circuit of capital and the subprime mortgage crisis. *International Journal of Urban and Regional Research* 33(2): 355–371.

Graham N. (2011). *Lawscape: Property, Environment, Law*. New York, NY: Routledge.

Gramsci, A. (1971). *Selection from the Prison Notebooks*. London: Lawrence and Wishart.

Granovetter, M. (1985). Economic action and social structure: The problem of embeddedness. *American Journal of Sociology* 91: 481–510.

Gutzon Larsen, H. and Lund Hansen, A. (2008). Gentrification – Gentle or traumatic? Urban renewal policies and socioeconomic transformations in Copenhagen. *Urban Studies* 45(12): 2429–2448.

Hackworth J. and Smith, N. (2001). The changing state of gentrification. *Tijdschrift voor Economische en Sociale Geografie* 92(4): 464–477.

Haila, A. (1990). The theory of land rent at the crossroads. *Environment and Planning D: Society and Space* 8(3): 275–296.

Haila, A. (2016). *Urban Land Rent. Singapore as a Property State*. Oxford: Wiley Blackwell.

Hale, H.E. (2011). Hybrid regimes: When democracy and autocracy mix. In: *Dynamics of Democratization* (ed. N. Brown), 23–45. Baltimore, MA: Johns Hopkins University Press.

Hall, P. (1986). *Governing the Economy. The Politics of State Intervention in Britain and France*. New York, NY and Oxford: Oxford University Press.

Hall, P. and Soskice, D. (2001). *Varieties of Capitalism. The Institutional Foundations of Comparative Advantage*. Oxford and New York, NY: Oxford University Press.

Hall, S. (1998). The great moving nowhere show. *Marxism Today*. Special edition (November/December): 9–14.

Hammel, D.J. (1999a). Re-establishing the rent gap: An alternative view of capitalised land rent. *Urban Studies* 36(8): 1283–1293.

Hammel, D.J. (1999b). Gentrification and land rent: A historical view of the rent gap in Minneapolis. *Urban Geography* 20(2): 116–145.

Hamnett, C. (1973). Improvement grants as an indicator of gentrification in Inner London. *Area* 5(4): 252–261.

Hamnett, C. (1976). Social change and residential segregation in Inner London 1961–1971. *Urban Studies* 13(3): 261–271.

Hamnett, C. (1991). The blind men and the elephant: The explanation of gentrification. *Transactions of the Institute of British Geographers* 16(2): 173–189.

Hamnett, C. (1992). Gentrifiers or lemmings? A response to Neil Smith. *Transactions of the Institute of British Geographers* 17(1): 116–119.

Hamnett, C. (2003). *Unequal City: London in the Global Arena*. London: Routledge.

Hamnett, C. (2009). Spatially displaced demand and the changing geography of house prices in London, 1995–2006. *Housing Studies* 24(3): 301–320.

Hamnett, C. (2010). Moving the poor out of London? The implications of the coalition government 2010 cuts to housing benefits. *Environment and Planning A* 42: 2809–2819.

Hamnett, C. and Randolph, B. (1988). *Cities, Housing and Profit: Flat Break-Up and the Decline of Private Renting*. London: Hutchinson.

Hamnett, C. and Randolph, W. (1984). The role of landlord disinvestment in housing market transformation: An analysis of the flat break-up market in Central London. *Transactions of the Institute of British Geographers* 9(3): 259–279.

Hamnett, C. and Williams, P. (1979). *Gentrification in London 1961–71. An Empirical and Theoretical Analysis of Social Change. (CURS Research Memorandum 71)*. Birmingham: Joint Centre for Regional, Urban and Local Government Studies.

Harré, R. (1986). *Varieties of Realism*. Oxford: Blackwell.

Harvey, D. (1973). *Social Justice and the City*. London: Edward Arnold.

Harvey, D. (1974). Class-monopoly rent, finance capital and the urban revolution. *Regional Studies* 8(3–4): 239–255.

Harvey, D. (1982). *The Limits to Capital*. Chicago, IL: University of Chicago Press.

Harvey, D. (1985). *The Urbanization of Capital. Studies in the History and. Theory of Capitalist Urbanization*. Baltimore, MA: Johns Hopkins University Press.

Harvey, B. (1994). Europe's homeless people and the role of housing. In: *Housing – Social integration and exclusion* (eds. H. Kristensen and J. Quigley). Horsholm: Danish Building Research Institute.

Harvey, D. (2010). *The Enigma of Capital and the Crisis of Capitalism*. London: Profile Books.

Häußermann, H. (1995). Von der 'sozialistischen' zur 'kapitalistischen' Stadt [From the 'socialist' to the 'capitalist' city]. *Aus Politik und Zeitgeschichte B* 12: 3–15.

Häußermann, H. and Kapphan, A. (2002). *Berlin: Von der geteilten zur gespaltenen Stadt? Sozialräumlicher Wandel seit 1990 [Berlin: From the Divided to the Separated City. Socio-Spatial Change Since 1990]*. Wiesbaden: Springer VS.

Häußermann, H., Holm, A., and Zunzer, D. (2002). *Stadterneuerung in der Berliner Republik [Urban Renewal in the New Republic]*. Wiesbaden: VS-Verlag.

He, S. (2019), Three waves of state-led gentrification in China. *Tijdschrift voor economische en sociale Geografie* 110(1): 26–34.

Healey, P. and Barrett, S. (1990). Structure and agency in land and property development processes: Some ideas for research. *Urban Studies* 27(1): 89–103.

Heller, M.A. (1999). The boundaries of property. *Yale Law Journal* 108(5): 1163–1223.

Heller, M.A. (1998). The tragedy of the anticommons. Property in the transition from Marx to markets. *Harvard Law Review* 111(3): 621–688.

Helms, H.G. (1992). *Die Stadt als Gabentisch. Beobachtungen der aktuellen Städtebauentwicklung [The City as a Table for Presents. Observation in Recent Urban Developments]*. Leipzig: Reclam.

Heywood, A. (2012). London for sale? An Assessment of the Private Housing Market in London and the Impact of Growing Overseas Investment. http://www.smith-institute.org.uk/wp-content/uploads/2015/10/London-for-sale.pdf (accessed 6 January 2022).

Hjödestrand, T. (2009). *Needed by Nobody: Homelessness, Humiliation and Humanness in Post-Socialist Russia*. Ithaca, NY: Cornell University Press.

HM Government. (2011). *Laying the Foundations: A Housing Strategy for England*. London: DETR. https://assets.publishing.service.gov.uk/government/uploads/system/uploads/attachment_data/file/7532/2033676.pdf (accessed 6 January 2022).

Hodkinson, S. (2011a). Housing regeneration and the private finance initiative in England: Unstitching the neoliberal urban straitjacket. *Antipode* 43: 358–383.

Hodkinson, S. (2011b). The private finance initiative in English council housing regeneration: A privatisation too far? *Housing studies* 26(6): 911–932.

Hodkinson, S. (2012). The new urban enclosure. *City* 16: 500–518.

Hodkinson, S. and Robbins, G. (2013). The return of class war conservatism? Housing under the UK Coalition government. *Critical Social Policy* 33(1): 57–77.

Hodkinson, S., Watt, P., and Mooney, G. (2013). Neoliberal housing policy – Time for a critical re-appraisal. *Critical Social Policy* 33(1): 3–16.

Holm A. (2006). *Die Restrukturierung des Raumes: Machtverhältnisse in der Stadterneuerung der 90er Jahre in Ostberlin [The Restructuring of Space: Power Relations in the Urban Renewal of the 1990s in East Berlin]*. Bielefeld: Transcript-Verlag.

Holm, A. (2010). Townhouses, Urban Village, Car Loft. Berliner Luxuswohnanlagen als 'dritte Welle' der Gentrification [Townhouses, urban village, car Loft. Berlin's luxury housing as a 'third wave' of gentrification]. *Geographische Zeitschrift* 98(2): 100–115.

Holm, A., Hamann, U,. and Kaltenborn, S. (2016). *Die Legende vom Sozialen Wohnungsbau*. Berlin: Berliner Hefte zu Geschichte und Gegenwart der Stadt.

Holmans, A.E. (1987). *Housing Policy in Britain*. London: Croom Helm.

Honoré, A.M. (1961). Ownership. In: *Oxford Essays in Jurisprudence* (ed. A.G. Guest), 107–147. Oxford: Oxford University Press.

Hope, A., Kato, Y., and J. Sbicca (2020). *A Recipe for Gentrification. Food, Power, and Resistance in the City*. New York, NY: New York University Press.

Housing Committee of St. Petersburg. (2014). *Gosudarstvennaya programma Sankt-Peterburga: Obespechenyie dostupnym zhilyem I zhilishzno-kommunalnymi usluga-mi zhiteley Sankt-Peterburga na 2015–2020* [State Programme for the Provision of Sufficient Housing and Housing Services for the Citizens of St. Petersburg 2015–2020. Proposal to the Resolution of the City of St. Petersburg No. 84, 18.2.2014]. http://46.182.26.58/SpbGovSearch/Document/7101.pdf (accessed 30 June 2017; webpage no longer available).

Imrie, R. and Raco, M. (2003). Community and the changing nature of urban policy. In: *Urban Renaissance? New Labour, Community and Urban Policy* (eds. R. Imrie and R. Mike), 3–36. Bristol: Policy Press.

Islam, T. and Sakızlıoğlu, B. (2015) The making of, and resistance to, state-led gentrification in Istanbul, Turkey. In: *Global Gentrifications* (eds. L. Lees, H.B. Shin, and E. López-Morales), 245–264. Bristol and Chicago, IL: Policy Press.

Jackson, B. (2012). Property-owning democracy: A short history. In: *Property-Owning Democracy: Rawls and Beyond* (eds. M. O'Neill and T. Williamson), 33–52. Oxford: Wiley Blackwell.

Jaedicke, W. and Wollmann, H. (1991). Wohnungspolitik und Regierungswechsel [Housing policies and changes of government]. In: *Die alte Bundesrepublik. Kontinuität und Wandel [The old Federal Republic of Germany. Continuity and change]*. Leviathan. Special Issue 12/1991 (eds. B. Blanke and H. Wollmann), 420–436. Opladen: Westdeutscher Verlag.

Jäger, J. (2003). Urban land rent theory: A regulationist perspective. *International Journal of Urban and Regional Research* 27(2): 233–249.

Jessop, B. (1990). *State Theory. Putting the Capitalist State in Its Place*. Cambridge: Polity Press.

Jessop, B. (2002). *The Future of the Capitalist State*. Cambridge: Polity Press.

Jessop, B. (2015). *The State: Past, Present and Future*. Cambridge: Polity Press.

Jones, C. and Murie, A. (2006). *The Right to Buy. Analysis & Evaluation of a Housing Policy*. Oxford: Blackwell.

Kallin, H. and Slater, T. (2014). Activating territorial stigma: Gentrifying marginality on Edinburgh's 'other' fringe. *Environment and Planning A* 46: 1351–1368.

Kary, K. (1988). The gentrification of Toronto and the rent gap theory. In: *The Changing Canadian Inner City* (eds. T. Bunting and P. Filion), 53–72. Waterloo: Department of Geography, University of Waterloo.

Katz, L. (2008). Exclusion and exclusivity in property law. *University of Toronto Law Journal* 58(3): 275–315.

Kemeny, J. (1995). *From Public Housing to the Social Market*. London: Routledge.

Kemeny, J. (2005). 'The really big trade off' between home ownership and welfare: Castles' evaluation of the 1980 thesis, and a reformulation 25 years on. *Housing, Theory and Society* 22(2): 59–75.

Kemeny, J. (2006). Corporatism and housing regimes. *Housing, Theory and Society* 23(1): 1–18.

Kemeny, J and Lowe, S. (1998). Schools of comparative housing research. From convergence to divergence. *Housing Studies* 13(2): 161–176.

Kemp, P. (1997). *A Comparative Study of Housing Allowance*. London: Department of Social Security.

King, P. (2001). *Understanding Housing Finance*. London: Routledge.

Kivell, P. (1993). *Land and the City*. London: Routledge

Knight Frank. Residential Research (2014). The Rental Revolution. Examining the Private Rented Sector 2014. http://my.knightfrank.co.uk/research-reports/private-rented-sector-report-2168.aspx (accessed 30 June 2014; webpage no longer available).

Kofner, St. (2004). *Wohnungsmarkt und Wohnungswirtschaft [Housing Market and Housing Economy]*. Munich: De Gruyter Oldenbourg.

Kollegova, A. (2011). Changes in Russian Housing Policy after the Crisis 2008. Paper presented at the New Housing Researchers' Day, July 4, 2011, European Network of Housing Researchers, Toulouse.

Kononenko, V. and Moshes A. eds. (2011). *Russia as a Network State. What Works in Russia When State Institutions Do Not?* New York, NY: Palgrave Macmillan.

Kornev, N. (2004). Segregatsya po-peterburgskyi – poysky realnosty [*Segregation in St. Petersburg – The retrieval of realities*]. *Pereskop* N.2/2004. http://teleskop-journal.spb.ru/files/dir_1/article_content1197390715339144file.pdf (accessed 30 June 2017).

Kosareva, N.B. and Struyk, R. (1993). Housing privatization in the Russian Federation. *Housing Policy Debate* 4(1): 81–100.

Krätke, S. (1995). *Stadt – Raum – Ökonomie, Einführung in aktuelle Problemfelder der Stadtökonomie und Wirtschaftsgeographie [City-Space-Economy. An Introduction into Current Problems of Urban and Economic Geographies]*. Basel; Boston, MA; and Berlin: Birkhäuser.

Krijnen, M. (2018). Gentrification and the creation and formation of rent gaps. *City* 22(3): 437–446.

Kuhnert, J. and Leps, O. (2017). *Neue Wohnungsgemeinnützigkeit. Wege zu langfristig preiswertem und zukunftsgerechtem Wohnraum [New Limited-Profit Housing. Ways Towards Long-Term Affordable and Sustainable Housing]*. Wiesbaden: Springer VS.

Lambert, C. (2002). Transforming the city: Post-recession gentrification and re-urbanisation. *XVI AESOP Congress*, Volos, Greece, 10–15 July 2002. http://eprints.uwe.ac.uk/9594 (accessed 23 November 2017).

Lapavitsas, C. (2013). *Profiting Without Producing: How Finance Exploits Us All*. New York, NY: Verso.

Lasswell, H.D. (1936). *Politics: Who Gets What, When, How*. New York, NY: Whittlesey House.

Ledeneva, A. (2006). *How Russia Really Works: The Informal Practices that Shaped Post-Soviet Politics and Business*. Ithaca, NY: Cornell University Press.

Lees, L. (1994). Gentrification in London and New York: An Atlantic gap? *Housing Studies* 9(2): 199–217.

Lees, L. (2003). Super-gentrification: The case of Brooklyn Heights, New York City. *Urban Studies* 40(12): 2487–2509.

Lees, L. (2008). Gentrification and social mixing: Towards an inclusive urban renaissance? *Urban Studies* 45(12): 2449–2470.

Lees, L. (2014). The urban injustices of New Labour's 'new urban renewal': The case of the Aylesbury Estate in London. *Antipode* 46(4): 921–947.

Lees, L. and Ferreri, M. (2016). Resisting gentrification on its final frontiers: Lessons from the Heygate Estate in London (1974–2013). *Cities* 57: 14–24.

Lees, L., Shin, H.B., and López-Morales, E. (eds.) (2015) *Global Gentrifications: Uneven Development and Displacement*. Bristol: Policy Press

Lees, L., Shin, H.B., and Lopez-Morales, E. (2016): *Planetary Gentrification*. Cambridge: Polity Press.

Lees, L., Slater, T., and Wyly, E. (2008). *Gentrification*. New York, NY: Routledge.

Lehmbruch, G. (1976). *Parteienwettbewerb im Bundesstaat [Party Competition in the Federal State]*. Stuttgart: Kohlhammer.

Lehmbruch, G. (1996). Die korporative Verhandlungsdemokratie in Westmitteleuropa [The corporative negotiating democracy in Western Central Europe]. *Swiss Political Science Review* 2: 1–24.

Lemanski C. (2014). Hybrid gentrification in South Africa: Theorising across southern and northern cities. *Urban Studies* 51(14): 2943–2960.

Levitas, R. (1998). *The Inclusive Society? Social Exclusion and New Labour*. Basingstoke: Palgrave Macmillan.

Ley, D. (1986). Alternative explanations for inner-city gentrification: A Canadian assessment. *Annals of the Association of American Geographers* 76(4): 521–535.

Ley, D. (1996). *The New Middle Class and the Remaking of the Central City*. Oxford: Oxford University Press.

Lijphart, A. (1971). Comparative politics and the comparative method. *American Political Science Review* 65 (September): 682–693.

Lijphart, A. (1977). *Democracy in Plural Societies: A Comparative Exploration*. New Haven, CT: Yale University Press.

Lijphart, A. (1999). *Patterns of Democracy*. New Haven, CT: Yale University Press.

Limonov L.E. (2011). Land market and urban development in Russia. In: *Regional Development and Policy – Challenges, Choices and Recipients* (ed. Regional Studies Association). Newcastle: Annual International Conference of the Regional Studies Association: 102–103.

Lind, H. (2001). Rent regulation: A conceptual and comparative analysis. *International Journal of Housing Policy* 1(1): 41–57.

London Assembly, Housing and Regeneration Committee (2013). Rent Reform. Making London's Private Rented Sector Fit for Purpose June 2013. https://www.london.gov.uk/about-us/london-assembly/london-assembly-publications/rent-reform-making-londons-private-rented (accessed 6 January 2022).

López-Morales, E. (2011). Gentrification by ground rent dispossession: The shadows cast by large-scale urban renewal in Santiago de Chile. *International Journal of Urban and Regional Research* 35: 330–357.

López-Morales, E., Shin, H.B., and Lees, L. (2016) Latin American gentrifications. *Urban Geography*, 37(8): 1091–1108.

Lowe, S. (2004). *Housing Policy Analysis. British Housing in Cultural and Comparative Context*. Basingstoke: Palgrave Macmillan.

Lowe, S. (2011). *The Housing Debate*. Bristol: The Policy Press.

Lowndes, V. and Pratchett, L. (2012). Local governance under the coalition government: Austerity, localism and the 'big society.' *Local Government Studies* 38(1): 21–40.

Lundquist, L. (1992). *Dislodging the Welfare State? Housing and Privatisation in Four European Countries*. Delft: Delft University Press.

Makhrova, A.G. and Golubchikov, O. (2012). Rossiyski gorod v uslovyach kapitalz-ma: sotsyalnaya transformatzya vnutrygorodskogo prostranstva [The Russian city under the conditions of capitalism: Social transformation and the transformation of the urban environment]. *Vestnik Moskovskogo Universiteta. Seria 5: Geografya* 2: 26–31.

Maloutas, T. (2007). Segregation, social polarization and immigration in Athens during the 1990s: Theoretical expectations and contextual difference. *International Journal of Urban and Regional Research* 31(4): 733–758.

Maloutas, T. (2012). Contextual diversity in gentrification research. *Critical Sociology* 38(1): 33–48.

Maloutas, T. (2018). Travelling concepts and universal particularisms: A reappraisal of gentrification's global reach. *European Urban and Regional Studies* 25(3): 250–265.

Malpass, P. (2003). The wobbly pillar? Housing policy and the British postwar welfare state. *Journal of Social Policy* 32(4): 589–606.

Malpass, P. (2005). *Housing & the Welfare State. The Development of Housing Policy in Britain*. Houndsmill: Palgrave Macmillan.

Malpass, P. and Victory, C. (2010). The modernisation of social housing in England. *International Journal of Housing Policy* 10(1): 3–18.

Marcińczak, S., Gentile, M., and Stępniak, M. (2013). Paradoxes of (post)socialist segregation: Metropolitan sociospatial divisions under socialism and after in Poland. *Urban Geography* 34(3): 327–352.

Marcuse, P. (1986). Abandonment, gentrification, and displacement: The linkages in New York City. In: *Gentrification of the City* (eds. N. Smith and P. Williams), 153–177. London: Unwin Hyman.

Marshall, T.H. (1950). *Citizenship and Social Class: And Other Essays*. Cambridge: University Press.

Marx, K. (1971 [1894]). *Capital, Volume III*. Moscow: Progress Publishers.

Massey, D. and Catalano, A. (1978). *Capital and Land, Landownership by Capital in Great Britain*. London: Edward Arnold.

Mau, St. (2003). *The Moral Economy of Welfare States: Britain and Germany Compared*. London: Routledge.

Mayer, A. (1998). *Theorie und Politik des Wohnungsmarktes [Theory and Policy of the Housing Market]*. Berlin: Duncker & Humblot.

McFarlane, C. (2010). The comparative city: Knowledge, learning, urbanism. *International Journal of Urban and Regional Research* 34(4): 725–742.

Merrett, S. (1979). *State Housing in Britain.* London: Routledge & Kegan Paul.

Merrett, S. and Gray, F. (1982). *Owner Occupation in Britain.* London: Routledge & Kegan Paul.

Merrill, T.W. and Smith, H.E. (2001). What happened to property in law and economics? *The Yale Law Journal* 111(2): 357–398.

Mints, V. (2000). Selecting a housing finance system for Russia. *Housing Finance International* 15: 49–57.

Mints, V. (2004). Strategies of transitional countries to establish a housing finance system. *The Russian experience. Housing Finance International* 19(2): 25–37.

Mitchell, D. (2003). *The Right to the City: Social Justice and the Fight for Public Space.* New York, NY: Guilford Press.

Modan, G.G. (2007). *Turf Wars, Discourse, Diversity, and the Politics of Place.* Malden, MA: Blackwell.

Nikonov, P. (2012). Narusheniya ochranogo zakonodetyelstva. XXI wek [Breaking conservation laws. 21st century]. In: *Sankt Peterburg: nasledye pod ugrozoy [St. Petersburg: Heritage at Risk]* (eds. C. Cecil and E. Minchenok), 157–163. Moscow: MAPS.

Novy, K. (1990). Gewinner und Verlierer im Kampf um Deregulierungsrenten. Skizze zur wohnungspolitischen Lage in der Bundesrepublik [Winners and losers in the struggle for deregulation-rents. A sketch on the situation of housing policies in Germany]. In: *Soziale Wohnungspolitik in den 90er Jahren. Probleme und Handlungsansätze aus britisch-deutscher Sicht [Social Housing Policy in the 90s. Problems and Approaches from a British-German Perspective]* (eds. K. Novy and A. Norton), 29–46. Basel: Birkhäuser.

O'Sullivan, D. (2002). Toward micro-scale spatial modeling of gentrification. *Journal of Geographical Systems* 4(3): 251–274.

Paccoud, A. (2016). Buy-to-let gentrification: Extending social change through tenure shifts. *Environment and Planning A* 49(4): 839–856.

Park, J. (2014). Land rent theory revisited. *Science & Society* 78(1): 88–109.

Patterson, D. (1996). *A Companion to Philosophy of Law and Legal Theory.* Oxford: Blackwell.

Pawson, H. and Mullins, D. (2010). *After Council Housing. Britain's New Social Landlords.* Houndsmill: Palgrave Macmillan.

Penner, J.E. (2009). Property, community, and the problem of distributive justice. *Theoretical Inquiries in Law* 10(1): 194–216.

Petrov, N., Lipman, M., and Hale, H.E. (2014). Three dilemmas of hybrid regime governance: Russia 848 from Putin to Putin. *Post-Soviet Affairs* 30(1): 1–26.

Phillips, M. (1993). Rural gentrification and the process of class colonization. *Journal of Rural Studies* 9(2): 123–140.

Phillips, M. (2004). Other geographies of gentrification. *Progress in Human Geography* 28(1): 5–30.

Pitt, J. (1977). *Gentrification in Islington. The Return of the Middle Classes to an Inner London Borough and the Subsequent Effects on Housing.* London: Barnsbury People Forum.

Polanyi, K. (1957 [1944]). *The Great Transformation: The Political and Economic Origins of Our Time*. Boston, MA: Beacon Press.

Porter L. and Shaw K. (2009). *Whose Urban Renaissance? An International Comparison of Urban Regeneration Strategies*. London: Routledge.

Porter, M. (2012). The rent gap at the metropolitan scale: New York City's land value valleys, 1990–2006. *Urban Geography* 31(3): 385–405.

Poulantzas, N. (2000 [1978]). *State, Power, Socialism*. London: Verso.

Power, A. (1972). *A Battle Lost. Barnsbury*. London: Friends House.

Power, A. (1973). *David and Goliath. Barnsbury 1973*. London: Holloway Law Centre and the Barnsbury Forum on Developments in Barnsbury.

Przeworski, A. and Teune, H. (1970). *The Logic of Comparative Social Inquiry*. New York, NY: Wiley.

Putnam, R.D. (1993). *Making Democracy Work: Civic Traditions in Modern Italy*. Princeton, NJ: Princeton University Press.

Putnam, R.D. (2000). *Bowling Alone: The Collapse and Revival of American Community*. New York, NY: Simon & Schuster.

Puzanov, A. (2012). Russia: The persistence of the socialist legacy. In: *Social Housing in Transition Countries* (eds. J. Hegedus, M. Lux, and N. Teller), 225–241. London: Routledge.

Ragin, C. (1987). *The Comparative Method. Moving Beyond Qualitative and Quantitative Strategies*. Berkeley, CA: University of California Press.

Ravetz, A. (2001). *Council Housing and Culture. The History of a Social Experiment*. London: Routledge.

Reimann B. (2000). *Städtische Wohnquartiere [Urban Residential Quarters]*. Opladen: Leske & Budrich.

Reulecke, J. (1985). *Geschichte der Urbanisierung in Deutschland [The History of Urbanization in Germany]*. Frankfurt a.M.: Suhrkamp.

Robinson, J. (2006). *Ordinary Cities*. London: Routledge.

Robinson, J. (2016). Thinking cities through elsewhere: Comparative tactics for a more global urban studies. *Progress in Human Geography* 40(1): 3–29.

Rolnik, R. (2013). Late neoliberalism: The financialization of homeownership and housing rights. *International Journal of Urban and Regional Research* 37(3): 1058–1066.

Rose, C.M. (1994). *Property as Persuasion: Essays, on the History, Theory, and Rhetoric of Ownership*. Boulder, CO: Westview Press.

Sakwa, R. (2010). The dual state in Russia. *Post-Soviet Affairs* 26(3): 185–206.

Sakwa, R. (2011). *The Crisis of Russian Democracy. The Dual State, Factionalism and the Medvedev Succession*. Cambridge: Cambridge University Press.

Sartori, G. (1970). Concept misformation in comparative politics. *The American Political Science Review* 64(4): 1033–1053.

Sartori, G. (1991). Comparing and miscomparing. *Journal of Theoretical Politics* 3(3): 243–257.

Savills World Research, UK Residential. (2013). Spotlight. The World in London. http://pdf.euro.savills.co.uk/residential---other/spot-worldlondon-lr.pdf (accessed 14 June 2014).

Savini, F. and Aalbers, M.B. (2016). The de-contextualisation of land use planning through financialisation: Urban redevelopment in Milan. *European Urban and Regional Studies* 23(4): 878–894.

Sayer, A. (1992). *Method in Social Science. A Realist Approach*. London: Routledge.

Sayer, A. (2000). *Realism and Social Science*. London: Sage.

Scanlon, K., Fernández, M., and Whitehead, C. (2014). *A Lifestyle Choice for Families? Private Renting in London, New York, Berlin and the Randstad*. London: Get Living London.

Scanlon, K. and Whitehead, C. ed. (2008). *Social Housing in Europe II – A Review of Policies and Outcomes*. London: London School of Economics.

Scharpf, F.W. (1985). Die Politikverflechtungs-Falle: Europäische Integration und deutscher Föderalismus im Vergleich [The joint decision-making trap. European integration and German federalism compared]. *Politische Vierteljahresschrift* 26: 323–356.

Schiebe, C. (2016). Verdrängung durch energetische Sanierung in Berlin [Displacement through energy-saving regeneration in Berlin]. https://blogs. hu-berlin.de/wohnenberlin/wp-content/uploads/sites/45/2016/08/Schiebe_ Verdr%C3%A4ngung-von-Bestandsmieterinnen.pdf (accessed 12 November 2017).

Schmid, C., Karaman, O., Hanakata, N.C. et al. (2018). Towards a new vocabulary of urbanisation processes: A comparative approach. *Urban Studies* 55(1): 19–52.

Schönig, B. (2013). Die neue Wohnungsfrage [The new housing question]. *Blätter für deutsche und internationale Politik* 2/2013: 17–20.

Schulz, G. (1986). Kontinuitäten und Brüche in der Wohnungspolitik von der Weimarer Zeit bis zur Bundesrepublik [Continuities and ruptures in housing policies from the Weimar Republic to the Federal Republic of Germany]. In: *Stadtwachstum, Industrialisierung, Sozialer Wandel. Beiträge zur Erforschung der Urbanisierung im 19. und 20. Jahrhundert [Urban Growth, Industrialization, Social Change. Contributions on Urbanization in the 19th and 20th Century]* (ed. H.J. Teuteberg), 135–174. Berlin: Duncker & Humblot.

Schwartz, H. and Seabrooke, L. (2008). Varieties of residential capitalism in the international political economy: Old welfare states and the new politics of housing. *Comparative European Politics* 6(3): 237–261.

Seeger, H. (1995). *Wohnungswirtschaft im Wahlzyklus der Politik [Housing Business and Electoral Cycles]*. Bad Sinzheim: Pro Universitate Verlag.

Senatsverwaltung für Bau- und Wohnungswesen von Berlin (1993). *Leitsätze zur Stadterneuerung in Berlin vom 31.August 1993* [Guiding Principles for Urban Renewal in Berlin as of 31 August 1993]. http://www.stadtentwicklung.berlin. de/staedtebau/foerderprogramme/stadterneuerung/de/download/leitsaetze.pdf (accessed 12 November 2017).

SEU. Social Exclusion Unit. (1998). *Bringing Britain Together. A National Strategy for Neighbourhood Renewal*. Cm. 4045, London: HMSO.

Sharafutdinova, G. (2010). *Political Consequences of Crony Capitalism Inside Russia*. Paris: Notre Dame University Press.

Shatkin, G. (2017). *Cities for Profit. The Real Estate Turn in Asia's Urban Politics*. Ithaca, NY: Cornell University Press.

Shevtsova, L. (2001). Ten years after the Soviet breakup: Russia's hybrid regime. *Journal of Democracy* 12(4): 65–70.

Shin, H.B. and Kim, S.H. (2016). The developmental state, speculative urbanization and the politics of displacement in gentrifying Seoul. *Urban Studies* 53(3): 540–559.

Shin, H.B. and López-Morales, E. (2018). Beyond Anglo-American gentrification theory. In: *Handbook of Gentrification Studies* (eds. L. Lees and M. Phillips), 13–25. Cheltenham: Edward Elgar.

Shin, H.B., Lees, L., and López-Morales, E. (2016) Introduction: Locating gentrification in the global East. *Urban Studies* 53(3): 455–47.

Shomina, E.S. (2010). *Kvartyrosemchyki – nashe 'zhilishnoye menshesto': rossyskiyi i zarubezhnyiv opyt razvitiiya arendnogo zhilya [Tenants – Our "Housing Minority:" Russian and Foreign Experiences in the Development of Rental Housing]*. Moscow: State University-Higher School of Economics Publishers.

Shomina, E.S. (2014). Wedomstvoye zhilye: Problemyi I perspektiviy [Departmental housing: Problems and perspectives]. *Munitsipalnaya wlast* 4–5: 84–92.

Singer, J.W. (2000). *Entitlement: The Paradoxes of Property*. New Haven, CT: Yale University Press.

Skocpol, T. (1979). *States and Social Revolutions. A Comparative Analysis of France, Russia and China*. New York, NY: Cambridge University Press.

Skocpol, T. and Somers, M. (1980). The uses of comparative history in macrosocial inquiry. *Comparative Studies in Society and History* 22(2): 174–197.

Slater, T. (2017). Planetary rent gaps. *Antipode* 49: 114–137.

Slater, T. (2018). Rent gaps. In: *Handbook of Gentrification Studies* (eds. L. Lees and M. Phillips), 119–133. Cheltenham: Edward Elgar.

Smart, A. and Smart, J. (2017). Ain't talkin' 'bout gentrification: The erasure of alternative idioms of displacement resulting from Anglo American academic hegemony. *International Journal of Urban and Regional Research* 41(3): 518–525.

Smelser, N. (1976). *Comparative Methods in the Social Sciences*. Englewood Cliffs, NJ: Prentice Hall.

Smelser, N. and Swedberg R. (2005). *The Handbook of Economic Sociology* (2nd expanded ed.). New York, NY and Princeton, NJ: Russell Sage Foundation and Princeton University Press.

Smith, D.P. and Holt, L. (2007). Studentification and 'apprentice' gentrifiers within Britain's provincial towns and cities: Extending the meaning of gentrification. *Environment and Planning A* 39(1): 142–161.

Smith, N. (1977). The return from the suburbs and the structuring of urban space: State involvement in Society Hill, Philadelphia. Unpublished BSc thesis. University of St Andrews.

Smith, N. (1979). Towards a theory of gentrification. A back to the city movement by capital, not people. *Journal of the American Planning Association* 45(4): 538–548.

Smith, N. (1982). Gentrification and uneven development. *Economic Geography* 58(2): 139–155.

Smith, N. (1984): *Uneven Development: Nature, Capital, and the Production of Space*. Oxford: Blackwell.

Smith, N. (1996). *The New Urban Frontier: Gentrification and the Revanchist City*. London: Routledge.

Smith, N. (2010). Toward a theory of gentrification. A back to the city movement by capital, not people . . . In: *The Gentrification Reader* (box 3) (eds. L. Lees, T. Slater, and E. Wyly), 97–98. New York, NY: Routledge.

Stedman Jones, D. (2012). *Masters of the Universe: Hayek, Friedman, and the Birth of Neoliberal Politics*, Princeton, NJ: Princeton University Press.

Stephens, M. and Whitehead, C. (2013). Rental housing policy in England: Post crisis adjustment or long term trend? *Journal of Housing and the Built Environment* 29: 201–220.

Stones, R. (1996). *Sociological Reasoning: Towards a Post Modern Sociology*. New York, NY: St. Martin's Press.

Struyk, R.J. and Kosareva, N.B. (1993). The Russian Housing Market in Transition (The Urban Institute Project Report). http://pdf.usaid.gov/pdf_docs/pnabq331.pdf (accessed 6 June 2017).

Swedberg, R. (2004). The toolkit for economic sociology. *CSES Working Paper Series Paper #22*, Cornell University.

Swenarton, M. (1981). *Homes Fit for Heroes. The Politics and Architecture of Early State Housing in Britain*, London: Routledge.

Sýkora, L. (1993). City in transition: The role of rent gaps in Prague's revitalization. *Tijdschjrift voor Economische en Sociale Geografie* 84(4): 281–293.

Szelenyi, I. (1983). *Urban Inequalities Under State Socialism*. New York, NY: Oxford University Press.

Szelenyi, I. (1996). Cities under socialism – and after. In: *Cities After Socialism* (eds. G. Andrusz, M. Harloe, and I. Szelenyi), 286–317. Oxford and Cambridge: Blackwell.

Tang, W. (2017). Beyond gentrification: Hegemonic redevelopment in Hong Kong. *International Journal of Urban and Regional Research* 41(3): 487–499.

Taşan-Kok, T. (2006). Institutional and spatial change. In: *The Urban Mosaic of Post-Socialist Europe: Space, Institutions, and Policy* (eds. S. Tsenkova and Z. Nedović-Budić), 51–70. Dordrecht: Springer.

Thomas, R. (2014). *What is the New 'Normal'? Mortgage Lending in 2014–15 and the March Back to a Sustainable Market* (Intermediary Mortgage Lenders Association Annual Report). http://www.imla.org.uk/perch/resources/imla-what-is-the-new-normal-mortgage-lending-in-2014-15-published.pdf (accessed 28 June 2014).

Tilly, C. (1984). *Big Structures, Large Processes, Huge Comparisons*. New York, NY: Sage.

Torgersen, U. (1987). Housing: The wobbly pillar under the welfare state. In: *Between State and Market: Housing in the Post-industrial Era* (eds. B. Turner, J. Kemeny, and L. Lundqvist), 116–126. Stockholm: Almqvist and Wicksell International.

Trumbull, N.S. (2012). Redevelopment and conservation in the built fabric of post-socialist St. Petersburg. *Urban Geography* 33(3): 370–400.

Turner, B. and Elsinga, M. (2005). Housing allowances: Finding a balance between social justice and market incentives. *International Journal of Housing Policy* 5(2): 103–109.

Tykhanova, E. and Kholova, A. (2015). Gorodskoy politicheskyj rezhym v Sankt-Peterburgye: rol realjnych I voobrozhayemych 'mashin rosta' v borbye za gorodskoye prostranstvo [Urban political regimes in St. Petersburg. The role of a real and a perceived 'growth machine" in struggles around the urban environment]. *Zhurnal isledovanyi sotsyjalnoy politiky* 13(2): 241–256.

UNUCE. United Nations, Economic Commission for Europe (2004). *Country Profile on the Housing Sector. Russian Federation*. Geneva: United Nations.

Valle, M.M. (2021), Globalizing the sociology of gentrification. *City & Community*. https://doi.org/10.1111/cico.12507.

van Gent, W. and Boterman, W. (2019). Gentrification of the changing state. *Tijdschjrift voor Economische en Sociale Geografie* 110(1): 35–46.

van Weesep, J. and Musterd, S. (1991). *Urban Housing for the Better-Off: Gentrification in Europe*. Utrecht: Stedelijke Netwerken.

Vesper, Dieter (1997). Berliner Finanzpolitik in einem fast ausweglosen Dilemma – aktuelle Tendenzen im Landeshaushalt von Berlin [Berlin's financial politics in an unsolvable dilemma. Recent tendencies in the state budget of Berlin]. *DIW Wochenbericht* 39/1997: 697–708.

Vesper, Dieter (1999). Kann sich Berlin aus seinem Finanzdilemma befreien? Zur Entwicklung des Landeshaushalts von Berlin [Can Berlin liberate itself from its financial dilemma? On the development of the state budget of Berlin]. *DIW Wochenbericht* 22: 389–397.

Vihavainen, R. (2009). *Homeowners' Associations in Russia – After the 2005 Housing Reform*. Helsinki: Kikimore Publications. http://www.iut.nu/members/Russia/HomeownerAssociations_Russia_2009.pdf (accessed 6 June 2016).

Vorobyev, D. and Campbell, T. (2009). Gentrification in Petersburg: A New City for a New Gentry. http://www.cisr.ru/files/publ/Vorobyev/berlin7.pdf (accessed 6 June 2016).

Walker, R. (1974). Urban ground rent: Building a new conceptual theory. *Antipode* 6(1): 51–58.

Wallace, A. (2016). *Remaking Community? New Labour and the Governance of Poor Neighbourhoods*. London: Routledge.

Ward, C. and Aalbers, M.B. (2016) 'The shitty rent business': What's the point of land rent theory? *Urban Studies* 53(9): 1760–1783.

Watt, P. (2001). *The Dynamics of Social Class and Housing: A Study of Local Authority Tenants in the London Borough of Camden*. Unpublished PhD Thesis. Department of Geography, King's College. London: University of London.

Watt, P. (2009a). Housing stock transfer, regeneration and state-led gentrification in London. *Urban Policy and Research* 27(3): 229–242.

Watt, P. (2009b). *Social housing and regeneration in London. In: Regenerating London. Governance, Sustainability and Community in a Global City* (eds. R. Imrie, L. Lees, and M. Raco), 212–236. London: Routledge.

Watt, P. (2013). 'It's not for us': Regeneration, the 2012 Olympics and the gentrification of East London. *City* 17(1): 99–118

Watt P. and Jacobs, K. (2000). Discourses of social exclusion. An analysis of bringing Britain together: A national strategy for neighbourhood renewal. *Housing, Theory and Society* 17(1): 14–26.

Watt, P. and Minton, A. (2016). London's housing crisis and its activisms. *City* 20(2): 204–221.

Welch Guerra, M. (2016). Mietpreisbindung als Dauerrecht [Rent control as a permanent right]. *Mieterecho* 384 (October16): 17–20.

Whitehead, C. (2012). Owner-occupation in an increasingly uncertain world: The English experience. In: *Beyond Home Ownership. Housing, Welfare and Society* (eds. R. Ronald and M. Elsinga), 108–129. Milton Park: Routledge.

Whitehead, C. and Scanlon, K. (2007). *Social Housing in Europe*. London: London School of Economics and Political Science.

Williams, P.R. (1976). The role of institutions in the Inner London housing market: The case of Islington. *Transactions of the Institute of British Geographers* 1(1): 72–82.

Williams, P.R. (1978). Building societies and the inner city. *Transactions of the Institute of British Geographers* 3(1): 23–34.

Wilson, J. and Johnson, J. (2013). The Future of London's Private Rented Sector (Future of London, Series Summary). https://www.landlordzone.co.uk/reports/future-of-londons-private-rented-sector.pdf (accessed 6 January 2022).

Winters, T. (1997). Erst Dinks, dann Kids? [First Dinks, then Kids?]. *VorOrt. Bauen und Wohnen in Prenzlauer Berg*.

Yeung, H.W. (1997). Critical realism and realist research in human geography: A method or a philosophy in search of a method? *Progress in Human Geography* 21(1): 51–74.

Yung, C.F. and King, R.J. (1998). Some tests for the rent gap theory. *Environment and Planning A* 30: 523–542.

Zavisca, J.R. (2008). Property without markets: Housing policy and politics in post-Soviet Russia, 1992–2007. *Comparative European Politics* 6: 365–386.

Zavisca, J.R. (2012). *Housing the New Russia*. Ithaca, NY: Cornell University Press.

Index

The Commodification Gap: Gentrification and Public Policy in London, Berlin and St Petersburg, First Edition. Matthias Bernt.
© 2022 John Wiley & Sons Ltd. Published 2022 by John Wiley & Sons Ltd.